Calculus

Differentiation & Integration

Lesson/Practice Workbook
for Self-Study and Test Preparation

Build Your Self-Confidence and Enjoyment of Math!
Comprehensive Solutions Manual Sold Separately

Aejeong Kang

MathRadar

Send all inquiries to:
MathRadar, LLC
5705 Spring Hill Dr.
Mckinney, Texas 75072

Visit www.mathradar.com for more information and a sneak preview of the MathRadar series of math books.

Send inquires via email: info@mathradar.com

Calculus (Differentiation & Integration): Lesson/Practice Workbook for Self-Study and Test Preparation

ISBN-13: 978-0-9893689-9-5

ISBN-10: 0989368998

Printed in the United States of America.

Preface

I wrote these books because I am a mother and I have a strong academic background in mathematics. I have a BS degree in Mathematics and Master's degree in Mathematics as well. I have completed Ph.D. program in Biostatistics.

After receiving the big blessing of our first child, a daughter, I decided to forgo my personal career goals to become a full-time mother. When our daughter entered 7th grade, that meant lots of help with her study of math-my passion. However, I struggled to find good math books that would help her understand difficult concepts both clearly and quickly. About two years ago, I talked with my husband and my kids (now I have 2 children 8th grader, Nichole and 1st grader, Richard) about an idea that it would be better to write math books myself at least for my kids because I really want my kids study math with best books. After the conversation, I decided that the best way to help my children was by writing math books for them myself. They wholeheartedly agreed.

That's why I've been able to pour all my knowledge, energy, and soul into Mathradar Series. Because I'm a mom, I would do anything for my children. Thanks to my family's endless support, I wrote them ten books, designed for use in junior high, high-school, and advanced high-school mathematics.

And that would have been the end of my journey, but my husband and children insisted that I share my work outside of our family. They encouraged me to make my work available to other parents looking, as I was, for well-written, great mathematics books for their children.

So I finally decided to publish these books. I do so with the hope that they will help your children find success and confidence in learning and studying mathematics.

But I would never have begun or finished this project without the support of my family. Kyungwan, Nichole, and Richard, you are my world. Thank you.

Aejeong Kang

Introduction

After reading several pages of explanation/description about a certain mathematical concept, you still don't get it.

You have worked on many related problems to understand mathematical concepts, but you still feel completely lost in the mathematical jungle.

You bought a math book with good reviews, but it only offers short answers without detailed solutions. You feel confused and frustrated.

You've tried multiple learning math books, but you've still not getting good grades in math. It seems like math is just not for you.

If any one of these situation sound familiar, the MathRadar series will help you escape!

Everyone has different learning abilities and academic skill. MathRadar series is written and organized with emphasis on helping each individual study mathematics at his/her own pace. Each book consists of clean and concise summaries, callouts, additional supporting explanations, quick reminders and/or shortcuts to facilitate better understanding. Each concept is thoroughly explained with step-by-step instruction and detailed proofs.
With the numerous examples and exercises, students can check their comprehension levels with both basic and more advanced problems.

Carry the MathRadar series with you!
Work on them anytime and anywhere!
Finally, you can start to enjoy mathematics!

Whether you are struggling or advanced in your math skills, the MathRadar series books will build your self-confidence and enjoyment of math.

I hope Math Radar is what you need and will be a great tool for your hard work.
Your comments or suggestions are greatly appreciated.
Please visit my website at www. mathradar.com or email me at ae-jeong@mathradar.com
Thank you very much. And remember, math can be fun!

TABLE OF CONTENTS

Chapter 1.   The Concept of Limits

Chapter 2. Limits of Functions and Continuity

Chapter 4.   Applications of the Derivative

Chapter 5.   The Indefinite Integral

Chapter 6.   The Definite Integral

Chapter 7.   Applications of the Integral

Answer Key

Index

Chapter 1. The Concept of Limits

1-1 Limits of Sequences

1. A Sequence of Real numbers

Note that there is a one-to-one correspondence between real numbers and points on a line.

Consider a finite sequence which can be described as an ordered array of the elements of a finite set.

For example, (1) $1, 2, 3, \cdots\cdots, n$

$$(2) \; 1, \frac{1}{2}, \frac{1}{3}, \cdots\cdots, \frac{1}{n}$$

For any finite sequence, there must be a first term and a last term.

An infinite sequence has a first term, a second term, a third term, etc., but no last term.

An infinite sequence is an ordered set of numbers in one-to-one correspondence with the natural numbers (positive integers) whose n^{th} term is $a_n = f(n)$, where f is a function.

Any mapping of the natural numbers into the real numbers is called a *sequence* of real numbers.

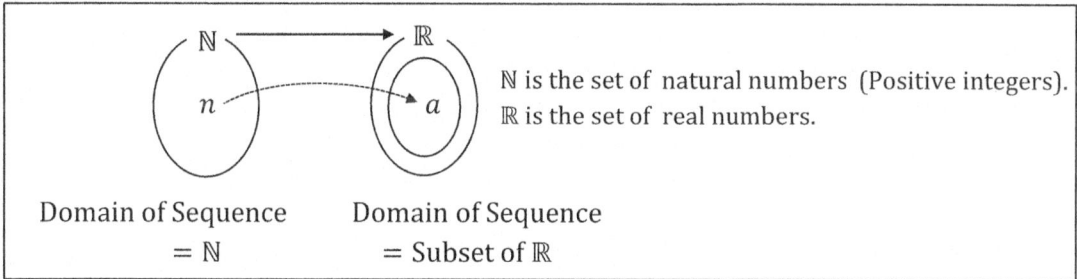

$\mathbb{N}$ is the set of natural numbers (Positive integers).
$\mathbb{R}$ is the set of real numbers.

Domain of Sequence = $\mathbb{N}$

Domain of Sequence = Subset of $\mathbb{R}$

For example, $\left\{(n, a) \middle| n \in \mathbb{N} \text{ and } a = \frac{1}{10^n}\right\} = \left\{0.1, 0.01, 0.001, \cdots\cdots, \frac{1}{10^n}, \cdots\cdots\right\}$,

$\{(n, a) | n \in \mathbb{N} \text{ and } a = (-1)^n\} = \{-1, 1, -1, \cdots\cdots, (-1)^n, \cdots\cdots\}$,

$\left\{(n, a) \middle| n \in \mathbb{N} \text{ and } a = \left(-\frac{1}{2}\right)^{n-1}\right\} = \left\{1, -\frac{1}{2}, \frac{1}{4} \cdots\cdots, \left(-\frac{1}{2}\right)^{n-1}, \cdots\cdots\right\}, \cdots\cdots$

It is convenient to abbreviate the name, $\{a_1, a_2, a_3, a_4, \cdots\cdots, a_n, \cdots\cdots\}$, by writing only the nth term of the sequence concerned, thereby obtaining $\{a_n\}$. That is, $\{(n, a) | n \in \mathbb{N} \text{ and } a = n^2\} = \{n^2\}$.

2. The Limit of a Sequence

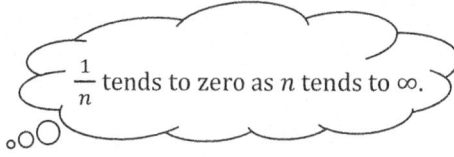

$\frac{1}{n}$ tends to zero as n tends to ∞.

Consider a sequence $\left\{1, \frac{1}{2}, \frac{1}{3}, \cdots\cdots, \frac{1}{n}, \cdots\cdots\right\} = \left\{\frac{1}{n}\right\}$.

For the sequence $\left\{\frac{1}{n}\right\}$, the terms are getting smaller and are getting closer and closer to zero.

That is, $\frac{1}{n}$ can be made as small as we pleased by choosing n sufficiently large.

Thus, $\left\{\frac{1}{n}\right\}$ is a *sequence of approximations* to zero.

(1) Definition

1) The sequence $\{a_n\}$ is a *sequence of approximations* to L if and only if $\lim\limits_{n \to \infty} a_n = L$

2) The sequence $\{a_n\}$ is said to be *convergent* if and only if there is a real number, say L,

   such that $\lim\limits_{n \to \infty} a_n = L$

   For example, $\quad \lim\limits_{n \to \infty} a_n = \lim\limits_{n \to \infty} \dfrac{1}{n} = 0, \quad \lim\limits_{n \to \infty} a_n = \lim\limits_{n \to \infty} \left(-\dfrac{1}{2}\right)^{n-1} = 0,$

   $$(a, a, a, \cdots\cdots, a); \ \lim\limits_{n \to \infty} a_n = \lim\limits_{n \to \infty} a = a$$

3) For a convergent sequence $\{a_n\}$,

   any real number L such that $\lim\limits_{n \to \infty} a_n = L$ is called the *limit of the sequence* $\{a_n\}$.

(2) Convergent Sequence

Consider a small positive number which we shall call ε.

For every $\varepsilon > 0$, there exists an integer N (depending on ε) for which $|a_n - L| < \varepsilon$ if $n > N$.

$\Leftrightarrow \ \lim\limits_{n \to \infty} a_n = L$

$\Leftrightarrow \ a_n \to L$ as $n \to \infty$

For a certain sequence $\{a_1, \ a_2, \ a_3, \ a_4, \ \cdots\cdots, a_n, \cdots\cdots\}$,

i) If $\lim\limits_{n \to \infty} a_n$ exists, we say the sequence converges. $\left(\lim\limits_{n \to \infty} a_n = L\right)$

ii) If the sequence does not converge, we say it diverges. $\left(\lim\limits_{n \to \infty} a_n = \infty \ \text{or} \ \lim\limits_{n \to \infty} a_n = -\infty\right)$

For example,

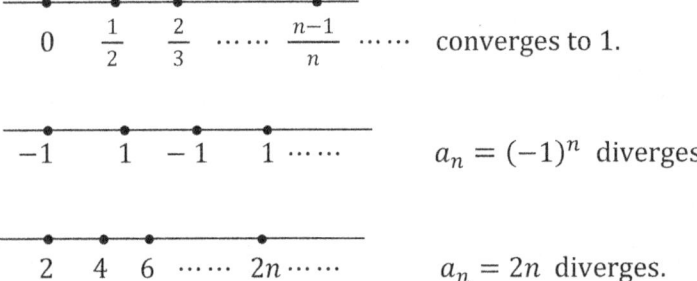

$$0 \quad \frac{1}{2} \quad \frac{2}{3} \quad \cdots\cdots \quad \frac{n-1}{n} \quad \cdots\cdots \quad \text{converges to 1.}$$

$$-1 \quad 1 \quad -1 \quad 1 \ \cdots\cdots \qquad a_n = (-1)^n \ \text{diverges.}$$

$$2 \quad 4 \quad 6 \ \cdots\cdots \ 2n \cdots\cdots \qquad a_n = 2n \ \text{diverges.}$$

(3) Uniqueness of the Limit

> Every convergent sequence has exactly one limit.

> The limit L for a given convergent sequence $\{a_n\}$ is unique.
>
> That is, $\quad$ if $\lim\limits_{n \to \infty} a_n = L$ and $L \ne M$, $\quad$ then $\lim\limits_{n \to \infty} a_n \ne M$

($\because$ Suppose $\lim\limits_{n \to \infty} a_n = M$.

Then, for every positive number ε, there exits an integer n_1 such that $|a_n - L| < \dfrac{\varepsilon}{2}$ when $n > n_1$.

Similarly, for the same ε, there exits an integer n_2 such that $|a_n - L| < \dfrac{\varepsilon}{2}$ when $n > n_2$.

Let n_3 be the larger of n_1 and n_2.

Since $|a_n - L| = |L - a_n|$ and the absolute value of a sum is equal to or less than the sum of the absolute values, $|L - M| = |(L - a_n) + (a_n - M)| \leq |L - a_n| + |a_n - M|$

$$= |a_n - L| + |a_n - M|$$

$$< \frac{\varepsilon}{2} + \frac{\varepsilon}{2} = \varepsilon \quad \text{for all } n > n_3$$

i. e., $|L - M| < \varepsilon$ for every arbitrary small number ε.

$\therefore L = M$, which contradicts the assumption.

Therefore, every convergent sequence has one and only one limit.)

3. Properties of Limits of Sequences

For convergent sequences $\{a_n\}, \{b_n\}$, let $\lim\limits_{n \to \infty} a_n = S$ and $\lim\limits_{n \to \infty} b_n = T$, and let k be a constant.

Then we have the following properties:

---

(1) $\lim\limits_{n \to \infty} ka_n = k \lim\limits_{n \to \infty} a_n = kS$

(2) $\lim\limits_{n \to \infty} (a_n \pm b_n) = \lim\limits_{n \to \infty} a_n \pm \lim\limits_{n \to \infty} b_n = S \pm T$

(3) $\lim\limits_{n \to \infty} a_n b_n = \lim\limits_{n \to \infty} a_n \cdot \lim\limits_{n \to \infty} b_n = ST$

(4) $\lim\limits_{n \to \infty} \dfrac{a_n}{b_n} = \dfrac{\lim\limits_{n \to \infty} a_n}{\lim\limits_{n \to \infty} b_n} = \dfrac{S}{T} \quad (b_n \neq 0, T \neq 0)$

Note :

(1) $\{a_n + b_n\}$ is convergent and $\lim\limits_{n \to \infty} (a_n \pm b_n) = \lim\limits_{n \to \infty} a_n \pm \lim\limits_{n \to \infty} b_n$,

   provided $\{a_n\}$ and $\{b_n\}$ are convergent.

(2) $\{a_n \cdot b_n\}$ is convergent and $\lim\limits_{n \to \infty} (a_n \cdot b_n) = \lim\limits_{n \to \infty} a_n \cdot \lim\limits_{n \to \infty} b_n$,

   provided $\{a_n\}$ and $\{b_n\}$ are convergent.

(3) $\left\{\dfrac{a_n}{b_n}\right\}$ is convergent and $\lim\limits_{n \to \infty} \dfrac{a_n}{b_n} = \dfrac{\lim\limits_{n \to \infty} a_n}{\lim\limits_{n \to \infty} b_n}$,

   provided $\{a_n\}$ and $\{b_n\}$ are convergent, $\lim\limits_{n \to \infty} b_n \neq 0$, and $b_n \neq 0$ for each n.

For example, consider the two convergent sequences $\{a_n\} = \{2\}$ and $\{b_n\} = \left\{\dfrac{1}{10^n}\right\}$ whose limits are 2 and 0, respectively.

Since the sequence $\left\{2 + \dfrac{1}{10^n}\right\}$ is the sum of the sequences $\{2\}$ and $\left\{\dfrac{1}{10^n}\right\}$,

$$\lim_{n\to\infty} \left(2 + \frac{1}{10^n}\right) = \lim_{n\to\infty} 2 \pm \lim_{n\to\infty} \frac{1}{10^n} = 2 + 0 = 2.$$

Similarly, the sequence $\left\{\frac{2}{10^n}\right\}$ can be expressed as the product of the sequences $\{2\}$ and $\left\{\frac{1}{10^n}\right\}$;

therefore, $\lim_{n\to\infty} \frac{2}{10^n} = \left(\lim_{n\to\infty} 2\right) \cdot \left(\lim_{n\to\infty} \frac{1}{10^n}\right) = 2 \cdot 0 = 0$

As another illustration, consider the convergent sequences $\left\{2 - \frac{1}{n}\right\}$ and $\left\{3 + \frac{2}{n}\right\}$ whose limits are

$$\lim_{n\to\infty} \left(2 - \frac{1}{n}\right) = \lim_{n\to\infty} 2 - \lim_{n\to\infty} \frac{1}{n} = 2 - 0 = 2 \quad ; \quad \lim_{n\to\infty} \left(3 + \frac{2}{n}\right) = \lim_{n\to\infty} 3 + \lim_{n\to\infty} \frac{2}{n} = 3 + 0 = 3$$

Note that the quotient of the sequences is the sequence $\left\{\frac{2 - \frac{1}{n}}{3 + \frac{2}{n}}\right\}$.

By the property, $\lim_{n\to\infty} \left(\frac{2 - \frac{1}{n}}{3 + \frac{2}{n}}\right) = \frac{\lim_{n\to\infty}\left(2 - \frac{1}{n}\right)}{\lim_{n\to\infty}\left(3 + \frac{2}{n}\right)} = \frac{2}{3}$

If convergent sequences $\{a_n\}$ and $\{b_n\}$ are not provided, $\underline{\lim_{n\to\infty} (a_n \cdot b_n) \neq \lim_{n\to\infty} a_n \cdot \lim_{n\to\infty} b_n}$

($\because$ If $a_n = n$ and $b_n = \frac{1}{n}$, then $\{a_n\}$ is divergent and $\{b_n\}$ is convergent.

Since $a_n b_n = 1$, $\lim_{n\to\infty} (a_n \cdot b_n) = 1$, $\lim_{n\to\infty} a_n = \infty$, and $\lim_{n\to\infty} b_n = 0$

Since $\lim_{n\to\infty} a_n = \infty$ and $\lim_{n\to\infty} b_n = 0$, $\lim_{n\to\infty} a_n \cdot \lim_{n\to\infty} b_n$ is not defined.

$\therefore$ $\lim_{n\to\infty} (a_n \cdot b_n) \neq \lim_{n\to\infty} a_n \cdot \lim_{n\to\infty} b_n$)

4. Computing the Limit of a Sequence

A sequence converges if and only if a limit exists as n approaches infinitely.

Compare the degree of the numerator and denominator to evaluate the limit.

(1) To evaluate the limits in the form $\frac{\infty}{\infty}$

First divide both numerator and denominator by the highest degree of denominator and use

$$\lim_{n\to\infty} \frac{1}{n} = 0 \quad ; \quad \lim_{n\to\infty} \frac{1}{n^2} = 0 \quad ; \quad \lim_{n\to\infty} \frac{1}{n^3} = 0$$

1) If the fraction contains a numerator and denominator of equal degree,

   then the limit is the quotient of their leading coefficients.

2) If the degree of the denominator is greater than the degree of its numerator,

   then the limit is 0.

3) If the degree of the numerator is greater than the degree of its denominator,

   then the limit does not exist; i.e., diverges to ∞ or $-\infty$.

Example

1) $\lim\limits_{n\to\infty} \dfrac{4n^2-3n}{3n^2+2n-1} = \lim\limits_{n\to\infty} \dfrac{4-\frac{3}{n}}{3+\frac{2}{n}-\frac{1}{n^2}} = \dfrac{4}{3}$; $\lim\limits_{n\to\infty} \dfrac{n(n-1)(2n-1)}{5n^3+2n} = \dfrac{2}{5}$

2) $\lim\limits_{n\to\infty} \dfrac{n^2-1}{n^3+2n} = \lim\limits_{n\to\infty} \dfrac{\frac{1}{n}-\frac{1}{n^3}}{1+\frac{2}{n^2}} = 0$

3) $\lim\limits_{n\to\infty} \dfrac{2n^3}{n^2+3} = \lim\limits_{n\to\infty} \dfrac{2n}{1+\frac{3}{n^2}} = \infty$ (No limit exists.)

---

(2) To evaluate the limits in the form $\infty - \infty$

---

1) If square roots are involved, then use rationalizing;

For example,

$$\lim_{n\to\infty}(\sqrt{n+1}-\sqrt{n}) = \lim_{n\to\infty}\frac{(\sqrt{n+1}-\sqrt{n})(\sqrt{n+1}+\sqrt{n})}{\sqrt{n+1}+\sqrt{n}} = \lim_{n\to\infty}\frac{n+1-n}{\sqrt{n+1}+\sqrt{n}} = \lim_{n\to\infty}\frac{1}{\sqrt{n+1}+\sqrt{n}} = 0$$

2) If square roots are not involved, then move the highest degree outside;

For example,

$$\lim_{n\to\infty}(3n^2-2n) = \lim_{n\to\infty}n^2\left(3-\frac{2}{n}\right) = \infty$$

5. Comparison about Limits

---

(1) If $\{a_n\}$ and $\{b_n\}$ are convergent sequences such that $a_n \le b_n$ for each n,

then $\lim\limits_{n\to\infty} a_n \le \lim\limits_{n\to\infty} b_n$

(2) If $\{a_n\}$ is convergent and $a_n \le b$ for each n, then $\lim\limits_{n\to\infty} a_n \le b$

(3) If $\{b_n\}$ is convergent and $a \le b_n$ for each n, then $a \le \lim\limits_{n\to\infty} b_n$

(4) If $a_n \le c_n \le b_n$ $(n = 1, 2, 3, \cdots)$ and $\lim\limits_{n\to\infty} a_n = \lim\limits_{n\to\infty} b_n = L$,

then $\lim\limits_{n\to\infty} c_n$ exists and is equal to L ; i.e., $\lim\limits_{n\to\infty} c_n = L$

---

1) If $a_n \ge 0$, then $\lim\limits_{n\to\infty} a_n \ge 0$

2) If $a_n \le b_n$ for each n, then

$\lim\limits_{n\to\infty} a_n - \lim\limits_{n\to\infty} b_n = \lim\limits_{n\to\infty}(a_n - b_n) \le 0$

$\therefore \lim\limits_{n\to\infty} a_n \le \lim\limits_{n\to\infty} b_n$

If $a_n = 1-\frac{1}{n}$ and $b_n = 1+\frac{1}{n}$, then
$a_n < b_n$ But, $\lim\limits_{n\to\infty} a_n = \lim\limits_{n\to\infty} b_n = 1$
That is, there are convergent sequences $\{a_n\}$ and $\{b_n\}$
such that $[a_n < b_n$, but $\lim\limits_{n\to\infty} a_n = \lim\limits_{n\to\infty} b_n]$

Example   Find the limit: $\lim\limits_{n\to\infty} \dfrac{1}{n}\sin\dfrac{n\pi}{2}$

Since $-1 \le \sin\dfrac{n\pi}{2} \le 1$, $\quad -\dfrac{1}{n} \le \dfrac{1}{n}\sin\dfrac{n\pi}{2} \le \dfrac{1}{n}$

Note that $\lim\limits_{n\to\infty} \dfrac{1}{n} = 0$; $\quad \lim\limits_{n\to\infty}\left(-\dfrac{1}{n}\right) = -\lim\limits_{n\to\infty}\dfrac{1}{n} = 0$

Therefore, $\lim\limits_{n\to\infty} \dfrac{1}{n}\sin\dfrac{n\pi}{2} = 0$

1-2 Limits of Geometric Sequences

Consider a geometric sequence $r, r^2, r^3, \cdots\cdots, r^{n-1}, r^n, \cdots\cdots$

$\boxed{\text{(1) When } r > 1, \quad \lim\limits_{n\to\infty} r^n = \infty \text{ (Diverge)}}$

For example, $\lim\limits_{n\to\infty} 2^n = \infty, \quad \lim\limits_{n\to\infty} 3^n = \infty$

$\boxed{\text{(2) When } r = 1, \quad \lim\limits_{n\to\infty} r^n = 1 \text{ (Converge)}}$

$\because \lim\limits_{n\to\infty} 1^n = \lim\limits_{n\to\infty} 1 = 1$

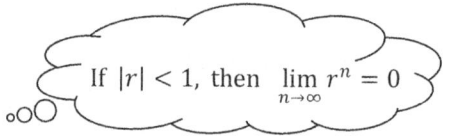

If $|r| < 1$, then $\lim\limits_{n\to\infty} r^n = 0$

$\boxed{\text{(3) When } -1 < r < 1, \quad \lim\limits_{n\to\infty} r^n = 0 \text{ (Converge)}}$

For example, $\lim\limits_{n\to\infty}\left(\dfrac{1}{2}\right)^n = 0, \quad \lim\limits_{n\to\infty} 0^n = 0, \quad \lim\limits_{n\to\infty}\left(-\dfrac{1}{2}\right)^n = 0$

$\boxed{\text{(4) When } r = -1, \quad \lim\limits_{n\to\infty} r^n \text{ diverges}}$

For example, $\lim\limits_{n\to\infty}(-1)^n$; $-1, 1, -1, 1, -1, \cdots\cdots$

$\boxed{\text{(5) When } r < -1, \quad \lim\limits_{n\to\infty} r^n \text{ diverges}}$

For example, $\lim\limits_{n\to\infty}(-2)^n$; $-2, 4, -8, 16, \cdots\cdots$

Note :

1) For a geometric sequence $\{r^n\}$ to have a limit, $-1 < r \le 1$.

2) For a geometric sequence $\{ar^{n-1}\}$ to have a limit, $a = 0$ or $-1 < r \le 1$.

Example   Determine whether the following sequences have limits.

In each case where a limit exists, calculate the limit.

$\quad$ (1) $\dfrac{\sqrt{2^n}+3^n}{3^n}$ $\qquad$ (2) $\dfrac{2^{2n}+3^{n+1}}{5^n+2^n}$ $\qquad$ (3) $2^n - 3^n$

(1) Divide both numerator and denominator by 3^n;

$$\lim\limits_{n\to\infty}\left(\frac{\sqrt{2^n}+3^n}{3^n}\right) = \lim\limits_{n\to\infty}\left\{\left(\frac{\sqrt{2}}{3}\right)^n + 1\right\} = 1 \text{ (The sequence converges to 1.)}$$

(2) Divide both numerator and denominator by 5^n;

$$\lim_{n\to\infty}\left(\frac{2^{2n}+3^{n+1}}{5^n+2^n}\right) = \lim_{n\to\infty}\left(\frac{4^n+3\cdot3^n}{5^n+2^n}\right) = \lim_{n\to\infty}\left(\frac{\left(\frac{4}{5}\right)^n+3\left(\frac{3}{5}\right)^n}{1+\left(\frac{2}{5}\right)^n}\right) = 0 \quad \text{(The sequence converges to 0.)}$$

(3) $\lim_{n\to\infty}(2^n-3^n) = \lim_{n\to\infty}3^n\left\{\left(\frac{2}{3}\right)^n-1\right\} = -\infty$ (The sequence diverges to $-\infty$.)

1-3 Series

1. Infinite Series

An expression of the form $a_1 + a_2 + a_3 + \cdots\cdots + a_n + \cdots\cdots$ is called an *infinite series*.

Each a_i is called a term and $a_1 + a_2 + a_3 + \cdots + a_n = \sum_{i=1}^{n} a_i$ is called the *partial sum* to n terms.

The sum of a finite geometric sequence is depending on the value of the common ratio r.

Now, consider the sum of an infinite geometric sequence.

For example, $\sum_{n=1}^{\infty}\left(\frac{1}{2}\right)^n = \frac{1}{2}+\frac{1}{4}+\frac{1}{8}+\frac{1}{16}+\cdots\cdots+\left(\frac{1}{2}\right)^n+\cdots\cdots$

Even though the series has infinitely many terms, it has a finite sum.

Let S_n be the partial finite sum of the first n terms.

Then, $S_n = \frac{\left(\frac{1}{2}\right)\left\{1-\left(\frac{1}{2}\right)^n\right\}}{1-\left(\frac{1}{2}\right)} = 1 - \left(\frac{1}{2}\right)^n$

$S_1 = a_1 = \frac{1}{2}$

$S_2 = a_1 + a_2 = \frac{1}{2}+\frac{1}{4} = \frac{3}{4}$

$S_3 = a_1 + a_2 + a_3 = \frac{1}{2}+\frac{1}{4}+\frac{1}{8} = \frac{7}{8}$

$\vdots$

$S_n = a_1 + a_2 + a_3 + \cdots\cdots + a_n = 1 - \left(\frac{1}{2}\right)^n$

$\therefore \lim_{n\to\infty}S_n = \lim_{n\to\infty}\left\{1-\left(\frac{1}{2}\right)^n\right\} = 1$

That is, S_n appears to be approaching 1 as n increases. ($S_n \to 1$ as $n \to \infty$)

In general, if the sequence of partial sums $S_1, S_2, \cdots\cdots, S_n, \cdots\cdots$ *converges* to S, then the series is said to converge to S, and S is called the sum of the infinite series; i.e., $\lim_{n\to\infty}S_n = S$.

Specially, $|r| < 1 \Rightarrow r^n$ becomes arbitrarily close to zero for n large.

Thus, we say the infinite geometric series

$a_1, a_1r, a_1r^2, a_1r^3, \cdots\cdots, a_1r^{n-1}, \cdots\cdots$ converges

$$S_n = \frac{a_1(1-r^n)}{1-r} \rightarrow \frac{a_1(1-0)}{1-r} = \frac{a_1}{1-r}$$

as $n \to \infty$

and it has the sum $S = \sum_{n=1}^{\infty} a_1 r^{n-1} = \frac{a_1}{1-r}$.

If $|r| \geq 1$, then the partial sum $S_n = a_1 + a_1 r + a_1 r^2 + \cdots \cdots + a_1 r^{n-1}$ is not arbitrarily close to a fixed number for n large. The sequence has no limit (no sum) and that the infinite series *diverges*.

The sum of an infinite series with first term a_1 and common ratio r is given by

---

(1) When $|r| < 1$, $S = \frac{a_1}{1-r}$ (Converge)

(2) When $|r| \geq 1$, the series has no sum. (Diverge)

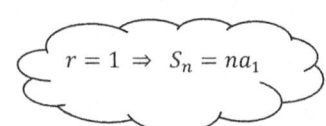

$r = 1 \Rightarrow S_n = na_1$

Example (1) Find the sum of the infinite geometric series $\sum_{i=1}^{\infty} 2\left(\frac{1}{5}\right)^{i-1}$.

(2) An infinite geometric series with first term 3 has a sum of 10.

Find the common ratio of the series.

(1) $a_1 = 2$, $r = \frac{1}{5}$

$\therefore S = \frac{a_1}{1-r} = \frac{2}{1-\frac{1}{5}} = \frac{2}{\frac{4}{5}} = \frac{10}{4} = \frac{5}{2}$

(2) $S = \frac{a_1}{1-r} = \frac{3}{1-r} = 10$

$\therefore 1 - r = \frac{3}{10}$ $\therefore r = \frac{7}{10}$ (Common ratio)

2. Limits of Infinite Series

Note that an infinite series *converges* if and only if the sequence of its partial sums converges to a finite, real number. If the limit does not exist, the series is said to *diverge*.

Consider the sum of an infinite geometric sequence.

For example, $1 + 3 + 5 + \cdots \cdots + (2n - 1) + \cdots \cdots$

Then, $S_n = \frac{n\{2 \cdot 1 + (n-1) \cdot 2\}}{2} = n^2$

$S_1 = a_1 = 1$

$S_2 = a_1 + a_2 = 1 + 3 = 4 = 2^2$

$S_3 = a_1 + a_2 + a_3 = 1 + 3 + 5 = 9 = 3^2$

$\vdots$

$S_n = a_1 + a_2 + a_3 + \cdots \cdots + a_n = n^2$

$\therefore \lim_{n \to \infty} S_n = \lim_{n \to \infty} n^2 = \infty$ (Diverge)

Note that $\boxed{\text{if } \sum_{n=1}^{\infty} a_n \text{ is convergent, then } \lim_{n \to \infty} a_n = 0}$

($\because$ Let $\sum_{n=1}^{\infty} a_n = S$ and S_n be the partial finite sum of the first n terms.

Then, $a_n = S_n - S_{n-1}$

$\therefore \lim_{n \to \infty} S_n = S$; $\lim_{n \to \infty} S_{n-1} = S$

$\therefore \lim_{n \to \infty} a_n = \lim_{n \to \infty} (S_n - S_{n-1}) = S - S = 0$

Therefore, $\lim_{n \to \infty} a_n = 0$)

If $\lim_{n \to \infty} a_n \neq 0$, then $\sum_{n=1}^{\infty} a_n$ is divergent.

However, $\boxed{\lim_{n \to \infty} a_n = 0 \not\Rightarrow \sum_{n=1}^{\infty} a_n \text{ is convergent.}}$

For example, let $a_n = \dfrac{1}{\sqrt{n+1}+\sqrt{n}}$

Then, $\lim_{n \to \infty} a_n = \lim_{n \to \infty} \dfrac{1}{\sqrt{n+1}+\sqrt{n}} = 0$

Let $S_n = \sum_{k=1}^{n} \dfrac{1}{\sqrt{k+1}+\sqrt{k}}$.

Then, $S_n = \sum_{k=1}^{n} (\sqrt{k+1} - \sqrt{k}) = (\sqrt{2} - 1) + (\sqrt{3} - \sqrt{2}) + (\sqrt{4} - \sqrt{3}) + \cdots\cdots + (\sqrt{n+1} - \sqrt{n})$

$$= -1 + \sqrt{n+1}$$

$\therefore \sum_{n=1}^{\infty} a_n = \sum_{n=1}^{\infty} \dfrac{1}{\sqrt{n+1}+\sqrt{n}} = \lim_{n \to \infty} S_n = \lim_{n \to \infty} (-1 + \sqrt{n+1}) = \infty$ (Diverge)

Example (1) Explain why the series $\sum_{n=1}^{\infty} \dfrac{2n}{n+1}$ diverges.

$\because$ Since $\lim_{n \to \infty} a_n = \lim_{n \to \infty} \dfrac{2n}{n+1} = 2 \neq 0$, the series $\sum_{n=1}^{\infty} \dfrac{2n}{n+1}$ diverges.

(2) Explain why the series $0.2 + 0.2^2 + 0.2^3 + \cdots\cdots$ converges.

$\because S_n = \dfrac{0.2\{1 - (0.2)^n\}}{1 - 0.2} = \dfrac{1}{4}\left\{1 - \left(\dfrac{1}{5}\right)^n\right\}$

$\therefore \lim_{n \to \infty} S_n = \lim_{n \to \infty} \left[\dfrac{1}{4}\left\{1 - \left(\dfrac{1}{5}\right)^n\right\}\right] = \dfrac{1}{4}$

$\therefore$ The series converges to $\dfrac{1}{4}$.

3. Telescoping Series and p-Series

(1) Telescoping Series

A *telescoping series* is a series whose partial sums eventually only have a fixed number of terms after cancellation of adjacent terms.

Consider the series which is the simplest example of a telescoping series.

$$\sum_{k=1}^{n} \frac{1}{k(k+1)} = \frac{1}{1\cdot2} + \frac{1}{2\cdot3} + \frac{1}{3\cdot4} + \frac{1}{4\cdot5} + \cdots + \frac{1}{n\cdot(n+1)}$$

Note that the *partial fraction decomposition*: $\boxed{\dfrac{1}{A\cdot B} = \dfrac{1}{B-A}\left(\dfrac{1}{A} - \dfrac{1}{B}\right)}$

Therefore, the series simplifies as

$$\sum_{k=1}^{n} \frac{1}{k(k+1)} = \sum_{k=1}^{n}\left(\frac{1}{k} - \frac{1}{k+1}\right) = \left(\frac{1}{1} - \frac{1}{2}\right) + \left(\frac{1}{2} - \frac{1}{3}\right) + \cdots + \left(\frac{1}{n-1} - \frac{1}{n}\right) + \left(\frac{1}{n} - \frac{1}{n+1}\right)$$

$$= 1 - \frac{1}{n+1} = \frac{n}{n+1}$$

Special Forms of the Partial Fraction Decomposition

1) $\dfrac{1}{n(n+a)} = \dfrac{1}{a}\left(\dfrac{1}{n} - \dfrac{1}{n+a}\right)$

2) $\dfrac{1}{(n+a)(n+b)} = \dfrac{1}{b-a}\left(\dfrac{1}{n+a} - \dfrac{1}{n+b}\right)$

3) $\dfrac{1}{n(n+1)(n+2)} = \dfrac{1}{2}\left(\dfrac{1}{n(n+1)} - \dfrac{1}{(n+1)(n+2)}\right)$

4) $\dfrac{1}{\sqrt{n+1}+\sqrt{n}} = \sqrt{n+1} - \sqrt{n}$

5) $\dfrac{1}{\sqrt{n+2}+\sqrt{n}} = \dfrac{1}{2}\left(\sqrt{n+2} - \sqrt{n}\right)$

6) $\log A + \log B = \log AB$; $\log A - \log B = \log\dfrac{A}{B}$

Example   Calculate the sum of the telescoping series: $\displaystyle\sum_{n=1}^{\infty} \frac{1}{n(n+2)}$

$$\sum_{n=1}^{\infty} \frac{1}{n(n+2)} = \sum_{n=1}^{\infty}\frac{1}{2}\left(\frac{1}{n} - \frac{1}{n+2}\right) = \frac{1}{2}\left\{\left(1 - \frac{1}{3}\right) + \left(\frac{1}{2} - \frac{1}{4}\right) + \left(\frac{1}{3} - \frac{1}{5}\right) + \cdots + \left(\frac{1}{n-1} - \frac{1}{n+1}\right) + \left(\frac{1}{n} - \frac{1}{n+2}\right) + \cdots\right\}$$

Except for 1 and $\dfrac{1}{2}$, all of the rational numbers in the series have corresponding opposites.

Therefore, $\displaystyle\sum_{n=1}^{\infty} \frac{1}{n(n+2)} = \frac{1}{2}\left(1 + \frac{1}{2}\right) = \frac{3}{4}$

(2) *p*-Series

p-series are infinite series of the form $\sum_{n=1}^{\infty} \frac{1}{n^p}$, where *p* is a positive real number.

When $p > 1$, the series converge.

When $0 < p \leq 1$, the series diverge.

Example   Determine the convergence of the series: $\sum_{n=1}^{\infty} \frac{1}{n^2}$

Since $\sum_{n=1}^{\infty} \frac{1}{n^2}$ is a *p*-series with $p = 2(> 1)$, $\sum_{n=1}^{\infty} \frac{1}{n^2}$ is convergent.

4. Geometric Series

Consider the (infinite) geometric series

$$1 + r + r^2 + r^3 + \cdots\cdots + r^n + \cdots\cdots$$

Or, generally, $a + ar + ar^2 + ar^3 + \cdots\cdots + ar^n + \cdots\cdots$ where $a \neq 0$

For a geometric sequence with first term *a* and common ratio *r*, the sum S_n of the first *n* terms is

$$S_n = a + ar + ar^2 + \cdots\cdots + ar^{n-1}$$

and $rS_n = ar + ar^2 + ar^3 + \cdots\cdots + ar^n$

By subtracting, we obtain

$$S_n - rS_n = a - ar^n \; ; \; (1-r)S_n = a(1-r^n)$$

$$\therefore \begin{cases} S_n = \frac{a(1-r^n)}{1-r} & , \; r \neq 1 \\ S_n = na & , \; r = 1 \end{cases}$$

Now, consider the limit of S_n.

If $|r| < 1$, then S_n has a limit.

Otherwise, S_n does not have a limit.

1) If $|r| < 1$, then $r^n \to 0$ as $n \to \infty$

$$\therefore \lim_{n \to \infty} S_n = \frac{a}{1-r}$$

2) If $|r| \geq 1$, then $r^n \to \infty$

$$\therefore \lim_{n \to \infty} S_n \text{ does not exist.}$$

The Sum of the Geometric Series

$$\lim_{n \to \infty} S_n = \lim_{n \to \infty} \frac{a(1-r^n)}{1-r} = \frac{a}{1-r} \; , \text{ provided } |r| < 1$$

$$\sum_{n=0}^{\infty} ar^n = \sum_{n=1}^{\infty} ar^{n-1} = a + ar + ar^2 + \cdots\cdots + ar^{n-1} + \cdots\cdots \quad (a \neq 0)$$

$\Rightarrow$ (1) If $|r| < 1$, then the series is convergent and

the sum of the series is $\sum_{n=1}^{\infty} ar^{n-1} = \frac{a}{1-r}$.

(2) If $|r| \geq 1$, then the series is divergent.

Example Determine the convergence of the series $\displaystyle\sum_{n=0}^{\infty} 5\left(\frac{2}{3}\right)^n$ and calculate the sum of the series.

Since $\displaystyle\sum_{n=0}^{\infty} 5\left(\frac{2}{3}\right)^n$ has the form $\displaystyle\sum_{n=0}^{\infty} ar^n$, it is a geometric series with $a = 5$, $r = \frac{2}{3}$

Note that the geometric series converges if $|r| < 1$.

Since $0 < \frac{2}{3} < 1$, the geometric series $\displaystyle\sum_{n=0}^{\infty} 5\left(\frac{2}{3}\right)^n$ converges.

Since the convergent geometric series $\displaystyle\sum_{n=0}^{\infty} ar^n$ has the sum $\frac{a}{1-r}$, the sum of the series $\displaystyle\sum_{n=0}^{\infty} 5\left(\frac{2}{3}\right)^n$ is

$$\sum_{n=0}^{\infty} 5\left(\frac{2}{3}\right)^n = \frac{a}{1-r} = \frac{5}{1-\frac{2}{3}} = \frac{5}{\frac{1}{3}} = 15$$

Exercises

#1 Determine whether the following sequences have limits.

(1) $\frac{2}{1}, \frac{2}{2}, \frac{2}{3}, \frac{2}{4}, \frac{2}{5}, \cdots\cdots$

(2) $0, \frac{1}{3}, \frac{2}{4}, \frac{3}{5}, \cdots\cdots$

(3) $\frac{1^2}{2}, \frac{2^2}{4}, \frac{3^2}{6}, \frac{4^2}{8}, \cdots\cdots$

(4) $\sin\frac{\pi}{2}, \sin\pi, \sin\frac{3\pi}{2}, \sin 2\pi, \cdots\cdots$

(5) $0.9, 0.99, 0.999, 0.9999, \cdots\cdots$

(6) $-1, -4, -9, -16, \cdots\cdots$

(7) $\{\log_2 n - \log_2(n+1)\}$

(8) $\{(-2)^{n-1}\}$

(9) $\{\sin\frac{n\pi}{2}\cos\frac{n\pi}{2}\}$

(10) $\{\tan(n\pi + (-1)^n \frac{\pi}{4})\}$

#2 Evaluate the following limits.

(1) $\lim\limits_{n\to\infty} \frac{3n-2}{5n+7}$

(2) $\lim\limits_{n\to\infty} \frac{6n^2+n+3}{5n^2-3n-4}$

(3) $\lim\limits_{n\to\infty} \frac{n^2+n-2}{4n^3-1}$

(4) $\lim\limits_{n\to\infty} \frac{2n^3}{n^2+1}$

(5) $\lim\limits_{n\to\infty} \{\log(n+3) - \log n\}$

(6) $\lim\limits_{n\to\infty} \frac{(n+1)(3n-2)}{(n-1)(2n+1)}$

(7) $\lim\limits_{n\to\infty} \frac{-2n^3-1}{n^2+1}$

(8) $\lim\limits_{n\to\infty} \{\log(10n^2-3n) - \log(n^2+2)\}$

(9) $\lim\limits_{n\to\infty} \left\{\left(1-\frac{1}{2^2}\right)\left(1-\frac{1}{3^2}\right)\left(1-\frac{1}{4^2}\right)\cdots\cdots\left(1-\frac{1}{n^2}\right)\right\}$

(10) $\lim\limits_{n\to\infty} \frac{1^2+2^2+3^2+\cdots\cdots+n^2}{n^3+1}$

(11) $\lim\limits_{n\to\infty} \frac{1\cdot2+2\cdot3+3\cdot4+\cdots\cdots+n(n+1)}{n(1+2+3+\cdots\cdots+n)}$

(12) $\lim\limits_{n\to\infty} (\sqrt{n+2} - \sqrt{n-2})$

(13) $\lim\limits_{n\to\infty} \frac{1}{\sqrt{n+1}-\sqrt{n}}$

(14) $\lim\limits_{n\to\infty} (\sqrt{n^2+n} - n)$

(15) $\lim\limits_{n\to\infty} (1 + 2n^2 - 3n^3)$

(16) $\lim\limits_{n\to\infty} (3n^3 - 2n^2 - n + 1)$

(17) $\lim\limits_{n \to \infty} \sqrt{n}(\sqrt{n+1} - \sqrt{n-1})$

(18) $\lim\limits_{n \to \infty} (\sqrt{1+2+3+\cdots\cdots+(n+1)} - \sqrt{1+2+3+\cdots\cdots+n})$

(19) $\lim\limits_{n \to \infty} \dfrac{a^{n+1}+b^{n+1}}{a^n+b^n}$ $(a > b > 0)$

(20) $\lim\limits_{n \to \infty} \dfrac{1+2+2^2+\cdots\cdots+2^n}{2^n}$

(21) $\lim\limits_{n \to \infty} (\sqrt{n^2+2n} + \sqrt{n^2+4n} + \cdots\cdots + \sqrt{n^2+20n} - 10n)$

(22) $\lim\limits_{n \to \infty} \dfrac{\sqrt{n^2-2014}-n}{n-\sqrt{n^2-2013}}$

(23) $\lim\limits_{n \to \infty} \dfrac{1^2-2^2+3^2-4^2+\cdots\cdots+(2n-1)^2-(2n)^2}{1-n^2}$

#3 Calculate the limits of the following sequences:

(1) $a_n = \dfrac{1}{1+r^n}$ $(r \neq -1)$

(2) $a_n = \dfrac{r^{n+1}-1}{r^n+1}$ $(r > 0)$

#4 Find the value.

(1) For two convergent sequences $\{a_n\}$ and $\{b_n\}$, $\lim\limits_{n \to \infty} (a_n + b_n) = 4$ and $\lim\limits_{n \to \infty} (a_n b_n) = 3$.

Find the value of $\lim\limits_{n \to \infty} a_n{}^2 + \lim\limits_{n \to \infty} b_n{}^2$.

(2) For a sequence $\{a_n\}$ such that $\lim\limits_{n \to \infty} (2n-1)\, a_n = 5$, find the value of $\lim\limits_{n \to \infty} n a_n$.

(3) For a sequence $\{a_n\}$ such that $\lim\limits_{n \to \infty} \dfrac{2a_n+3}{a_n-1} = 4$, find the value of $\lim\limits_{n \to \infty} a_n$.

(4) Find the value of n such that $n^{n^{n^{\cdot^{\cdot^{\cdot^n}}}}} = \dfrac{2}{3}$.

(5) For a positive integer n, $A_n = \left\{ (x,y) \,\Big|\, 0 \le x \le \dfrac{1}{2} + \dfrac{1}{2^2} + \dfrac{1}{2^3} + \cdots\cdots + \dfrac{1}{2^n},\ 0 \le y \le 1 \right\}$.

When a_n is the maximum value of $2x + y$, find the value of $\lim\limits_{n \to \infty} a_n$.

(6) For real numbers a, b and c such that $\lim\limits_{n \to \infty} \dfrac{bn^3+cn-4}{an^2+3n-2} = 5$, find the value of $a+b+c$.

(7) For real numbers a and b such that $\lim\limits_{n \to \infty} \left(\sqrt{n^2+an+3} - \sqrt{n^2+bn+2} \right) = 4$,

find the value of $a - b$.

(8) For convergent sequence $\{a_n\}$ such that $\lim\limits_{n \to \infty} \dfrac{3^n-2^{n+1}a_n}{3^n a_n+2^n} = 5$, find the value of $\lim\limits_{n \to \infty} a_n$.

(9) When an infinite geometric sequence $\left\{ \left(\dfrac{a}{2} \right)^{n-1} (a+2) \right\}$ is convergent,

find the range of the real number a.

(10) For any positive integer n, $A = \begin{bmatrix} n^2 + n & 2 \\ n^2 & 3 \end{bmatrix}$ is a 2×2 matrix.

When a_n is the sum of all entries of the inverse matrix of A, find the value of $\lim\limits_{n\to\infty} a_n$.

(11) For any positive integer n, $f(n) = 1 + 2 + 3 + \cdots\cdots + n$. Find the value of $\lim\limits_{n\to\infty} \dfrac{f(2n^2)}{\{f(n)\}^2}$.

(12) For two points $P(n, f(n))$ and $Q(n+1, f(n+1))$ on a quadratic function $f(x) = 2x^2$,

let a_n be the distance between P and Q.

Find the value of $\lim\limits_{n\to\infty} \dfrac{a_n}{n}$ (where n is a positive integer).

(13) For a sequence $\{a_n\}$, $a_n = n\left(\sqrt{\dfrac{n-1}{2n+1}} - a\right)$ ($n \geq 1$, n; positive integer),

find the value of the real number a so that $\lim\limits_{n\to\infty} a_n$ exists and calculate the limit.

(14) For sequences $\{a_n\}$ and $\{b_n\}$ such that $\lim\limits_{n\to\infty}(n-1)a_n = 5$ and $\lim\limits_{n\to\infty}(2n^2-1)b_n = 10$,

find the value of $\lim\limits_{n\to\infty}(n+1)^3 a_n b_n$.

(15) For a convergent sequence $\{a_n\}$, $\dfrac{a_n - 4}{a_n + 2} = \dfrac{1^3 + 2^3 + 3^3 + \cdots\cdots + n^3}{n^4 + 1}$. Find the value of $\lim\limits_{n\to\infty} a_n$.

(16) When $\lim\limits_{n\to\infty}(a_n - b_n) = 2$ and $\lim\limits_{n\to\infty} a_n = \infty$, find the value of $\lim\limits_{n\to\infty}\left(\dfrac{a_n^2}{b_n} - \dfrac{b_n^2}{a_n}\right)$. $(a_n b_n = 0)$

(17) For a sequence $\{a_n\}$ such that $[\log_2 a_n] = 3n$ (n; positive integer),

find the value of $\lim\limits_{n\to\infty} \dfrac{\log_8 a_n}{3n}$. (Where $[x]$ is the greatest integer less than or equal to x.)

(18) For real numbers a ($a < 0$) and b, $a + b = 1$.

For a matrix $A = \begin{bmatrix} 1 & a \\ 0 & b \end{bmatrix}$, $A^n = \begin{bmatrix} 1 & a_n \\ 0 & b_n \end{bmatrix}$ ($n = 1, 2, 3, \cdots$). Find the value of $\lim\limits_{n\to\infty} \dfrac{b_n}{a_n}$.

#5 Find the limit.

(1) For a sequence $\{a_n\}$ such that $a_1 = 1$ and $a_{n+1} = \dfrac{1}{2}a_n + 3$ ($n \geq 1$), find: 1) a_n 2) $\lim\limits_{n\to\infty} a_n$.

(2) For a sequence $\{a_n\}$ such that $a_1 = 1$ and $a_{n+1} = 2a_n$, find the value of $\lim\limits_{n\to\infty} \dfrac{a_n}{3^{n-1}}$.

(3) For a sequence $\{a_n\}$ such that $a_1 = 1$, $a_2 = 3$, and $2a_{n+2} - 3a_{n+1} + a_n = 0$ ($n \geq 1$),

find the value of $\lim\limits_{n\to\infty} a_n$.

(4) For a sequence $\{a_n\}$, let S_n be the partial finite sum of the first n terms.

Find the value of $\lim\limits_{n\to\infty} a_n$ 1) when $S_1 = 10$ and $S_{n+1} = \dfrac{1}{2}S_n + 1$ ($n = 1, 2, 3, \cdots\cdots$)

2) when $S_n = \dfrac{n(n+3)}{4(n+1)(n+2)}$

(5) For a positive integer n, let a_n be the decimal part of $\sqrt{n^2+3n}$. Find the value of $\lim\limits_{n\to\infty}\dfrac{10}{a_n}$.

(6) For a sequence $\{a_n\}$ such that $2n^2-1 < na_n < 2n^2+1$ $(n=1,2,3,\cdots\cdots)$,

find the value of $\lim\limits_{n\to\infty}\dfrac{a_n+3n-1}{n+3}$.

(7) For a convergent sequence $\{a_n\}$, $\lim\limits_{n\to\infty}\dfrac{a_{n-1}+4}{a_{n+1}-4}=5$. Find the value of $\lim\limits_{n\to\infty}a_n$.

(8) For a sequence $\{a_n\}$ such that $a_1=2$ and $a_{n+1}=\dfrac{1}{2}a_n+5$ $(n=1,2,3,\cdots\cdots)$,

find the value of $\lim\limits_{n\to\infty}a_n$.

(9) For a sequence $\{a_n\}$ (n; positive integer), $2n+1 < a_n < 2n+2$.

Find the value of $\lim\limits_{n\to\infty}\dfrac{n^2}{a_n+a_{n+1}+a_{n+2}+\cdots\cdots+a_{2n}}$.

(10) For a sequence $\{a_n\}$ such that $2n-1 < na_n-\cos n\theta < 2n+1$, find the value of $\lim\limits_{n\to\infty}a_n$.

#6 For sequences $\{a_n\}$ and $\{b_n\}$, determine whether the statement is true or false.

(1) If $a_n < b_n$ and $\lim\limits_{n\to\infty}a_n=\infty$, then $\lim\limits_{n\to\infty}b_n=\infty$.

(2) If $\{a_n\}$ and $\{b_n\}$ are convergent such that $a_n < b_n$, then $\lim\limits_{n\to\infty}a_n < \lim\limits_{n\to\infty}b_n$.

(3) If $\lim\limits_{n\to\infty}a_nb_n=0$, then $\lim\limits_{n\to\infty}a_n=0$ or $\lim\limits_{n\to\infty}b_n=0$.

(4) If $\{a_n\}$ is divergent and $\{a_n+b_n\}$ converges to 0, then $\{b_n\}$ is divergent.

(5) If $\{a_n-b_n\}$ converges to 0, then $\{a_n\}$ and $\{b_n\}$ are convergent.

(6) If $\left\{a_n+\dfrac{n}{2n+1}\right\}$ is convergent, then $\{a_n{}^2\}$ is convergent.

(7) For two convergent sequences $\{a_n\}$ and $\{b_n\}$,

if $\{a_n-b_n\}$ converges to a negative number, then $a_n < b_n$.

(8) If $\lim\limits_{n\to\infty}a_{2n-1}=a$, $\lim\limits_{n\to\infty}a_{2n}=b$, and $a < b$, then $\{a_n\}$ is convergent.

(9) For two convergent sequences $\{a_n\}$ and $\{b_n\}$,

if $a_n < c_n < b_n$ and $\lim\limits_{n\to\infty}(b_n-a_n)=0$, then $\{c_n\}$ is convergent.

(10) For a sequence $\{a_n\}$ with $a_1=1$, $\dfrac{1}{a_{n+1}}=\dfrac{1}{a_n}+1$ $(n=1,2,3,\cdots\cdots)$, $\lim\limits_{n\to\infty}a_n=0$

(11) If S_n is the finite sum of first n terms such that $S_n=ka_n+1$ $(k\,(\neq 1);\text{constant})$, then

i) $a_1=\dfrac{1}{1-k}$

ii) $\{a_n\}$ is a geometric sequence.

iii) If $k=\dfrac{2}{3}$, then $\{a_n\}$ is convergent.

#7 For a sequence $\{a_n\}$ with $a_1 = 2$ and $a_i > 0$ $(i = 1, 2, 3, \cdots\cdots)$,

a quadratic equation $x^2 - \sqrt{a_n}\, x + (a_{n+1} - 1) = 0$ has a double root for any positive integer n.

Find the value of $\lim\limits_{n \to \infty} a_n$.

#8 For a sequence $\{a_n\}$ with $a_n > 0$ $(n = 1, 2, 3, \cdots\cdots)$, an inequality $x^2 - 4\sqrt{2a_{n+1}}\, x + a_n > 0$ is always true.

    (1) Find the relationship between a_n and a_{n+1} .

    (2) Find the value of $\lim\limits_{n \to \infty} a_n$.

    (3) Find the value of $\lim\limits_{n \to \infty} \dfrac{3a_n + 2n - 1}{5a_n + 4n + 6}$.

#9 For a positive integer n, when a polynomial $f(x) = 2^n x^2 + 3^n x + 1$ is divided by $x - 1$ and $x - 2$, the remainders are a_n and b_n, respectively. Find the value of $\lim\limits_{n \to \infty} \dfrac{a_n}{b_n}$.

#10 Find the sum, when the sum exists.

    (1) $\displaystyle\sum_{n=2}^{\infty} \log \dfrac{n^2}{n^2 - 1}$

    (2) $\displaystyle\sum_{n=1}^{\infty} (\sqrt{n+2} - \sqrt{n+1})$

    (3) $\left(\dfrac{3}{2} - \dfrac{4}{3}\right) + \left(\dfrac{4}{3} - \dfrac{5}{4}\right) + \left(\dfrac{5}{4} - \dfrac{6}{5}\right) + \cdots\cdots$

    (4) $\log_2 \left(1 - \dfrac{1}{2^2}\right) + \log_2 \left(1 - \dfrac{1}{3^2}\right) + \cdots\cdots + \log_2 \left(1 - \dfrac{1}{(n+1)^2}\right) + \cdots\cdots$

    (5) $\displaystyle\sum_{n=1}^{\infty} \left(\dfrac{2}{3}\right)^n \sin\left(n\pi + \dfrac{\pi}{6}\right)$

#11 Determine the convergence of the following series.

    (1) $\displaystyle\sum_{n=1}^{\infty} 2n^{-\frac{3}{2}}$
          (2) $\displaystyle\sum_{n=1}^{\infty} \dfrac{\sqrt[4]{n}}{3\sqrt[3]{n^2}}$

#12 Determine the convergence of the following series. If the series converges, then calculate its sum.

    (1) $1 + \dfrac{1}{2} + \dfrac{1}{2^2} + \dfrac{1}{2^3} + \cdots\cdots$

    (2) $(\sqrt{3} + 1) + (\sqrt{3} - 3) + (9\sqrt{3} - 15) + \cdots\cdots$

(3) $\sin 30° + \sin^2 30° + \sin^3 30° + \sin^4 30° + \cdots\cdots$

(4) $\displaystyle\sum_{n=1}^{\infty} 2^n \left(\frac{1}{3}\right)^{n-1}$

(5) $\displaystyle\sum_{n=0}^{\infty} 3\left(-\frac{2}{5}\right)^n$

(6) $\displaystyle\sum_{n=1}^{\infty} \frac{2^n + (-3)^n}{4^n}$

(7) $\displaystyle\sum_{n=0}^{\infty} \left(3^{n+1} - 1\right)\left(\frac{1}{4}\right)^n$

(8) $\displaystyle\sum_{n=1}^{\infty} \left(\frac{1}{2}\right)^n \cos n\pi$

(9) $\displaystyle\sum_{n=1}^{\infty} \left(-\frac{1}{2}\right)^n \sin\left(n\pi + \frac{\pi}{3}\right)$

(10) $\displaystyle\lim_{k\to\infty} \frac{1}{k} \sum_{n=1}^{k} \left(\sum_{j=1}^{n} \frac{1}{2^j}\right)$

#13 Find the value.

(1) When a_n is the nth term of a sequence 2, 4, 8, 16, $\cdots\cdots$ and S_n is the finite sum of first n terms, find the value of $\displaystyle\lim_{n\to\infty} \frac{a_n}{S_n}$.

(2) For $f(x) = \frac{1}{x(x+1)}$,

  1) Find the minimum value of the integer n such that $\displaystyle\sum_{k=1}^{n} f(k) > 0.9$

  2) Find the value of positive constant a such that $\displaystyle\lim_{n\to\infty} \sum_{k=1}^{n} f(a + k) = \frac{1}{2}$

(3) For a quadratic equation $x^2 + (n - 1)x + n^2 = 0$,

let α_n and β_n be the two roots of the equation. Find the value of $\displaystyle\sum_{n=1}^{\infty} \frac{1}{(\alpha_n - 1)(\beta_n - 1)}$.

(4) For a quadratic equation $(4n^2 - 1)x^2 - 4nx + 1 = 0$ (n; positive integer),

let α_n and β_n ($\alpha_n > \beta_n$) be two roots of the equation. Find the value of $\displaystyle\sum_{n=1}^{\infty} (\alpha_n - \beta_n)$.

(5) For convergent series $(a_1 - 2) + (a_2 - \frac{3}{2}) + (a_3 - \frac{4}{3}) + (a_4 - \frac{5}{4}) + \cdots\cdots$,

find the value of $\displaystyle\lim_{n\to\infty} a_n$.

(6) For a sequence $\{a_n\}$ such that $\displaystyle\sum_{n=1}^{\infty} \left(a_n - \frac{2^{n+1} + 3^n}{3^{n+1}}\right) = 5$, find the value of $\displaystyle\lim_{n\to\infty} \frac{-a_n + 1}{3a_n - 2}$.

(7) When $a_n = \displaystyle\sum_{k=1}^{n} ak$, find the value of the constant a such that $\displaystyle\sum_{n=1}^{\infty} \frac{1}{a_n} = \frac{1}{2}$.

(8) For a geometric sequence $\{a_n\}$, $\displaystyle\sum_{n=1}^{\infty} a_n = 3$ and $\displaystyle\sum_{n=1}^{\infty} a_n^2 = 4$. Find the value of $\displaystyle\sum_{n=1}^{\infty} a_n^3$.

(9) For a sequence $\{a_n\}$, $a_1 = 1$, $a_2 = 2$, and $a_{n+2} = a_{n+1} + a_n$ $(n = 1, 2, 3, \cdots\cdots)$.

Find the value of $\displaystyle\sum_{n=1}^{\infty} \frac{a_n}{a_{n+1}a_{n+2}}$.

(10) For an arithmetic sequence $\{a_n\}$ with first term -100 and common difference 4,

a sequence $\{b_n\}$ is defined by $b_n = \displaystyle\sum_{k=1}^{n} (a_{k+1} - a_k)$. Find the value of $\displaystyle\lim_{n \to \infty} \sum_{k=1}^{n} \frac{1}{(2n+1)^2} b_k$.

(11) When an infinite series $S = 1 + \dfrac{1}{4} + \dfrac{1}{9} + \dfrac{1}{16} + \cdots\cdots + \dfrac{1}{n^2} + \cdots\cdots$ converges,

express the value of $1 + \dfrac{1}{9} + \dfrac{1}{25} + \dfrac{1}{49} + \cdots\cdots + \dfrac{1}{(2n-1)^2} + \cdots\cdots$ as S.

(12) Find the range of x so that $\displaystyle\sum_{n=1}^{\infty} \left(\frac{x}{2}\right)^n$ and $\displaystyle\sum_{n=1}^{\infty} \left(\frac{1}{x}\right)^n$ are convergent.

(13) Find the range of x so that the infinite geometric sequence $\{(3x-1)^{n-1}\}$ and the infinite

geometric series $\displaystyle\sum_{n=1}^{\infty} \left(\frac{\log_2 x - 1}{2}\right)^n$ are convergent.

(14) When $\displaystyle\sum_{n=1}^{\infty} \frac{a_n}{2^n} = 1$, find the value of $\displaystyle\lim_{n \to \infty} \frac{2^{n-1} + 5a_n}{2^{n+3} - 3a_n}$.

(15) For a sequence $\{a_n\}$, let S_n be the finite partial sum of first n terms such that $\displaystyle\lim_{n \to \infty} S_n = 3$.

Find the value of $\displaystyle\lim_{n \to \infty} \frac{a_n + S_n}{S_n{}^3}$.

(16) For a sequence a_1, $2a_2$, $2^2 a_3$, $\cdots\cdots$, $2^{n-1}a_n$, $\cdots\cdots$, the sum of first n terms is $10n$.

Find the value of $\displaystyle\sum_{n=1}^{\infty} a_n$.

(17) When $\cos^2 \theta + \cos^2 \theta \sin \theta + \cos^2 \theta \sin^2 \theta + \cdots\cdots = \dfrac{18}{13}$ $\left(0 < \theta < \dfrac{\pi}{2}\right)$,

find the value of $\dfrac{10}{\tan \theta}$.

(18) For the infinite sequence $\sqrt{3}$, $\sqrt{3\sqrt{3}}$, $\sqrt{3\sqrt{3\sqrt{3}}}$, $\cdots\cdots$ find the limit of the sequence.

(19) For a sequence $\{a_n\}$ with $a_1 = 4$ and $a_{n+1} = \sqrt[3]{a_n}$ $(n \geq 1)$,

find the value of $\displaystyle\lim_{n \to \infty} (a_1 \cdot a_2 \cdot a_3 \cdot \cdots\cdots \cdot a_n)$.

(20) For a geometric sequence $\{a_n\}$, the sum of the first term and second term is 4 and

the sum of the second term and third term is -2. Find the value of $\displaystyle\sum_{n=1}^{\infty} a_n$.

(21) When $S = 1 + \dfrac{1}{2} + \dfrac{1}{2^2} + \dfrac{1}{2^3} + \cdots\cdots$ and S_n is the finite sum of first n terms of the series,

find the minimum value of n such that $S - S_n \leq 0.01$ $(\log 2 = 0.3010)$

(22) For a sequence $\{a_n\}$ such that $\displaystyle\sum_{k=1}^{n} a_k = 3\left\{1 - \left(\frac{1}{3}\right)^n\right\}$, find the value of $\displaystyle\sum_{n=1}^{\infty} a_{2n}$.

(23) For a sequence $\{a_n\}$ with $a_1 = 1$ and $a_n = \sum_{k=1}^{n-1} a_k$ $(n \geq 2)$, find the value of $\sum_{n=1}^{\infty} \dfrac{1}{a_n}$.

(24) When the infinite series $\sum_{n=1}^{\infty} \dfrac{an^2 + 3}{n^2 + 2n}$ is convergent, find the sum of the series. $(a;$ constant$)$

#14 For sequences $\{a_n\}$ and $\{b_n\}$, determine if the following statements are true or false.

(1) If $\sum_{n=1}^{\infty} 2a_n$ converges, then $\lim_{n \to \infty} a_n = 0$.

(2) If $\lim_{n \to \infty} a_n = 0$, then $\sum_{n=1}^{\infty} a_n{}^2$ converges.

(3) If $\sum_{n=1}^{\infty} a_n b_n$ converges, then $\lim_{n \to \infty} a_n = 0$ and $\lim_{n \to \infty} b_n = 0$.

(4) If $\sum_{n=1}^{\infty} a_n$ and $\sum_{n=1}^{\infty} (a_n + b_n)$ converges, then $\sum_{n=1}^{\infty} b_n$ converges.

(5) If $\sum_{n=1}^{\infty} a_n b_n$ converges and $\lim_{n \to \infty} a_n \neq 0$, then $\lim_{n \to \infty} b_n = 0$.

(6) If $\sum_{n=1}^{\infty} a_n = a$ and $\sum_{n=1}^{\infty} b_n = b$, then $\sum_{n=1}^{\infty} a_n b_n = ab$. $(a, b;$ constants$)$

(7) If $\sum_{n=1}^{\infty} (2a_n + b_n)$ and $\sum_{n=1}^{\infty} (a_n - 2b_n)$ converge, then $\sum_{n=1}^{\infty} a_n$ and $\sum_{n=1}^{\infty} b_n$ converge.

(8) If $a_n > b_n$, $\sum_{n=1}^{\infty} a_n = a$, and $\sum_{n=1}^{\infty} b_n = b$, then $a > b$. $(a, b;$ real numbers, $n;$ positive integer$)$

(9) If $\sum_{n=1}^{\infty} a_{2n-1} = a$ and $\sum_{n=1}^{\infty} a_{2n} = b$, then $\sum_{n=1}^{\infty} a_n = a + b$.

#15 For geometric sequences $\{a_n\}$ and $\{b_n\}$, determine if the following statements are true or false.

(1) If $\sum_{n=1}^{\infty} a_n$ converges, then $\sum_{n=1}^{\infty} a_{2n}$ converges.

(2) If $\sum_{n=1}^{\infty} a_n$ diverges, then $\sum_{n=1}^{\infty} a_{2n}$ diverges.

(3) If $\sum_{n=1}^{\infty} a_n$ converges, then $\sum_{n=1}^{\infty} \left(a_n + \dfrac{1}{2}\right)$ converges.

(4) If $\sum_{n=1}^{\infty} a_n$ and $\sum_{n=1}^{\infty} b_n$ converge, then $\sum_{n=1}^{\infty} a_n b_n$ converges.

(5) If $\sum_{n=1}^{\infty} a_n b_n$ converges, then at least one of $\sum_{n=1}^{\infty} a_n$ and $\sum_{n=1}^{\infty} b_n$ is convergent.

(6) If $\displaystyle\sum_{n=1}^{\infty} a_n$ and $\displaystyle\sum_{n=1}^{\infty} b_n$ are divergent, then $\displaystyle\lim_{n \to \infty} (a_n + b_n) \neq 0$.

(7) If $\displaystyle\sum_{n=1}^{\infty} a_n{}^3$ and $\displaystyle\sum_{n=1}^{\infty} b_n{}^3$ are convergent, then $\displaystyle\sum_{n=1}^{\infty} (a_n + b_n)$ is convergent.

Chapter 2. Limits of Functions and Continuity

2-1 Limits of Functions

1. Introduction to Limits

(1) Intuitive Meaning of Limits

Consider the graph of the function $f(x) = x^2 + 1$.

Let P be a point moving along the x-axis, and

let Q be the point directly above P on the graph of f.

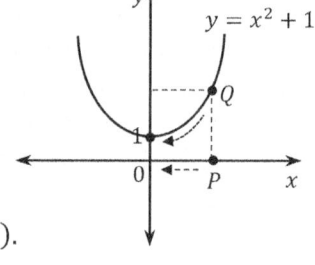

As P tends toward the origin, Q slides down the parabola toward $(0, 1)$.

That is, as x approaches 0, $f(x)$ approaches 1.

The notation is: $f(x) \to 1$ as $x \to 0$.

In limit notation, we write $\lim\limits_{x \to 0} f(x) = 1$.

This is read "The limit of $f(x)$ as x approaches 0 is 1."

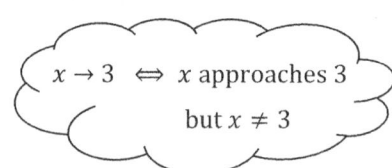

$x \to 3 \iff x$ approaches 3 but $x \neq 3$

Note:

$\lim\limits_{x \to a} f(x) = L$ means that when x is near but different from a, $f(x)$ is near L.

Example 1) Find $\lim\limits_{x \to 3}(4x - 2)$ 2) Find $\lim\limits_{x \to 3} \dfrac{x^2 - x - 6}{x - 3}$

1) When x is near 3, $4x - 2$ is near $4 \cdot 3 - 2 = 10$

   We write $\lim\limits_{x \to 3}(4x - 2) = 10$.

2) $\lim\limits_{x \to 3} \dfrac{x^2 - x - 6}{x - 3} = \lim\limits_{x \to 3} \dfrac{(x-3)(x+2)}{x - 3} = \lim\limits_{x \to 3}(x + 2) = 3 + 2 = 5$

(2) One-Sided Limits

Recall that $[x]$ denotes the greatest integer in x. ($[x]$ is the greatest integer less than or equal to x).

For any real number x and integer n, $n \leq x < n + 1 \iff [x] = n$

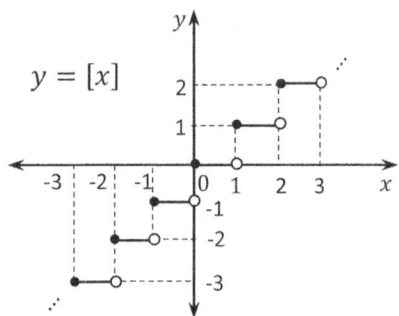

$y = [x]$

$-2 \leq x < -1 \implies [x] = -2 \qquad \therefore f(x) = -2$

$-1 \leq x < 0 \implies [x] = -1 \qquad \therefore f(x) = -1$

$0 \leq x < 1 \implies [x] = 0 \qquad \therefore f(x) = 0$

$1 \leq x < 2 \implies [x] = 1 \qquad \therefore f(x) = 1$

$2 \leq x < 3 \implies [x] = 2 \qquad \therefore f(x) = 2$

For all numbers x less than 2 but near 2, $[x] = 1$,

but for all numbers x greater than 2 but near 2, $[x] = 2$.

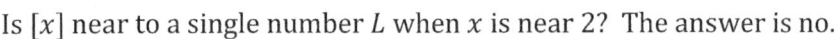

The graph of $[x]$ takes a jump at each integer.

Is $[x]$ near to a single number L when x is near 2? The answer is no.

Therefore, we conclude $\lim\limits_{x \to 2} [x]$ does not exist.

If a function takes a jump as does $[x]$ at each integer, then the limit does not exist at the jump points.

For such functions, we introduce one-sided limits.

$x \to a^+$ means that x approaches a from the right.

$x \to a^-$ means that x approaches a from the left.

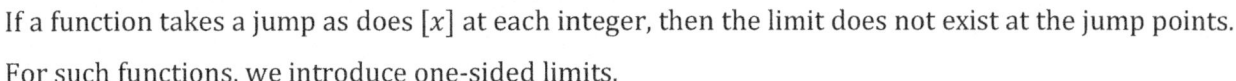

$\lim\limits_{x \to a^+} f(x) = L$; Limit from the right of $f(x)$ as x tends to a.

$\lim\limits_{x \to a^-} f(x) = L$; Limit from the left of $f(x)$ as x tends to a.

Right- and Left- Hand Limits

$\lim\limits_{x \to a^+} f(x) = L$ means that when x is near but on the right of a, $f(x)$ is near L.

Similarly, $\lim\limits_{x \to a^-} f(x) = L$ means that when x is near but on the left of a, $f(x)$ is near L.

$$\lim\limits_{x \to a} f(x) = L \underset{\text{if and only if}}{\Longleftrightarrow} \lim\limits_{x \to a^+} f(x) = \lim\limits_{x \to a^-} f(x) = L$$

That is, a necessary and sufficient condition that $\lim\limits_{x \to a} f(x)$ exist is that

both $\lim\limits_{x \to a^+} f(x)$ and $\lim\limits_{x \to a^-} f(x)$ exist and be equal.

Example Show that $\lim\limits_{x \to 2} [x]$ does not exist.

$\because$ Since $\lim\limits_{x \to 2^-} [x] = 1$ and $\lim\limits_{x \to 2^+} [x] = 2$, $\lim\limits_{x \to 2^-} [x] \neq \lim\limits_{x \to 2^+} [x]$

Therefore, $\lim\limits_{x \to 2} [x]$ does not exist.

Consider the function $f(x) = \dfrac{x}{|x|}, \ x \neq 0$

As x tends to 0 from the right, $f(x)$ tends to $+1$, but as x tends to 0 from the left, $f(x)$ tends to -1.

The notation for this situation is:

$\lim\limits_{x \to 0^+} \dfrac{x}{|x|} = \lim\limits_{x \to 0^+} \dfrac{x}{x} = 1$

$\lim\limits_{x \to 0^-} \dfrac{x}{|x|} = \lim\limits_{x \to 0^-} \dfrac{x}{(-x)} = -1$

$\therefore \ \lim\limits_{x \to 0^+} \dfrac{x}{|x|} \neq \lim\limits_{x \to 0^-} \dfrac{x}{|x|}$

Therefore, $\lim\limits_{x \to 0} \dfrac{x}{|x|}$ does not exist.

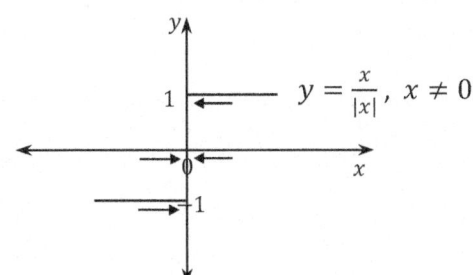

$y = \dfrac{x}{|x|}, \ x \neq 0$

(3) Limits and Infinity

Consider the graph of the function $f(x) = \dfrac{1}{x^2}$

$f(x)$ may grow large without bound as x approaches 0.

We say the limit is $+\infty$; $\displaystyle\lim_{x \to 0} f(x) = \infty$

Similarly, the function $f(x) = -\dfrac{1}{x^2}$ may decrease without bound as x approaches 0.

We say the limit is $-\infty$; $\displaystyle\lim_{x \to 0} f(x) = -\infty$

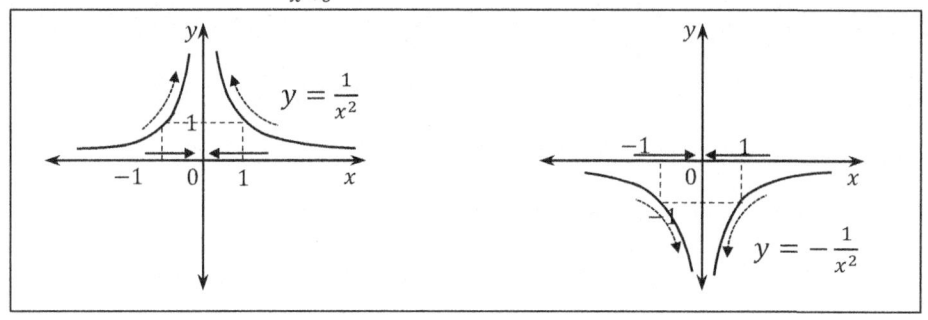

Example Evaluate the following limits:

$$(1)\ \lim_{x \to 0^+} \frac{1}{x} \qquad (2)\ \lim_{x \to 0^-} \frac{1}{x} \qquad (3)\ \lim_{x \to \infty^+} \frac{1}{x} \qquad (4)\ \lim_{x \to \infty^-} \frac{1}{x}$$

(1) $\displaystyle\lim_{x \to 0^+} \frac{1}{x} = \infty$

(2) $\displaystyle\lim_{x \to 0^-} \frac{1}{x} = -\infty$

(3) $\displaystyle\lim_{x \to \infty^+} \frac{1}{x} = 0$

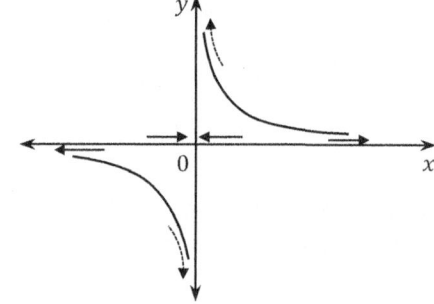

(4) $\displaystyle\lim_{x \to \infty^-} \frac{1}{x} = 0$

Since $\displaystyle\lim_{x \to 0^+} \frac{1}{x} \neq \lim_{x \to 0^-} \frac{1}{x}$, $\displaystyle\lim_{x \to 0} \frac{1}{x}$ does not exist.

2. Properties of Limits (Functions)

For a positive integer n and a constant k, let f and g be functions which have limits at a.

Then, we have the following properties:

(1) $\displaystyle\lim_{x \to a} k = k$

(2) $\displaystyle\lim_{x \to a} x = a$

(3) $\displaystyle\lim_{x \to a} kf(x) = k \lim_{x \to a} f(x)$

(4) $\displaystyle\lim_{x \to a} \{f(x) + g(x)\} = \lim_{x \to a} f(x) + \lim_{x \to a} g(x)$; $\displaystyle\lim_{x \to a} \{f(x) - g(x)\} = \lim_{x \to a} f(x) - \lim_{x \to a} g(x)$

(5) $\lim\limits_{x \to a} \{f(x) \cdot g(x)\} = \lim\limits_{x \to a} f(x) \cdot \lim\limits_{x \to a} g(x)$

(6) $\lim\limits_{x \to a} \dfrac{f(x)}{g(x)} = \dfrac{\lim\limits_{x \to a} f(x)}{\lim\limits_{x \to a} g(x)}$, provided $\lim\limits_{x \to a} g(x) \neq 0$

(7) $\lim\limits_{x \to a} \{f(x)\}^n = \{\lim\limits_{x \to a} f(x)\}^n$

(8) $\lim\limits_{x \to a} \sqrt[n]{f(x)} = \sqrt[n]{\lim\limits_{x \to a} f(x)}$, provided $\lim\limits_{x \to a} f(x) > 0$ when n is even

Substitution Theorem

If f is a polynomial function or a rational function, then $\lim\limits_{x \to a} f(x) = f(a)$

provided in the case of a rational function that the value of the denominator at $x = a$ is not zero.

Squeeze Theorem

Let f, g and h be functions such that $f(x) \leq g(x) \leq h(x)$ for all x near a, except possibly at a.

If $\lim\limits_{x \to a} f(x) = \lim\limits_{x \to a} h(x) = L$, then $\lim\limits_{x \to a} g(x) = L$

Note: When $f(x) < g(x)$, we may have $\lim\limits_{x \to a} f(x) = \lim\limits_{x \to a} g(x)$.

For example, let $f(x) = 3x$ and $g(x) = x^2 + 3x$ for $x \neq 0$.

Then, $f(x) < g(x)$

But, $\lim\limits_{x \to 0} f(x) = \lim\limits_{x \to 0} g(x) = 0$

Note: For two functions $f(x)$ and $g(x)$, let $\lim\limits_{x \to a} \dfrac{f(x)}{g(x)} = c$ (c is a constant). Then,

1) If $\lim\limits_{x \to a} g(x) = 0$, then $\lim\limits_{x \to a} f(x) = 0$

$\because \lim\limits_{x \to a} f(x) = \lim\limits_{x \to a} \left\{\dfrac{f(x)}{g(x)} \cdot g(x)\right\} = \lim\limits_{x \to a} \dfrac{f(x)}{g(x)} \cdot \lim\limits_{x \to a} g(x) = c \cdot 0 = 0$

2) If $\lim\limits_{x \to a} f(x) = 0$ $(a \neq 0)$, then $\lim\limits_{x \to a} g(x) = 0$

$\because \lim\limits_{x \to a} g(x) = \lim\limits_{x \to a} \left\{f(x) \div \dfrac{f(x)}{g(x)}\right\} = \lim\limits_{x \to a} f(x) \div \lim\limits_{x \to a} \dfrac{f(x)}{g(x)} = \dfrac{0}{c} = 0$

3) If $\lim\limits_{x \to a} f(x) = \infty$ $(a \neq 0)$, then $\lim\limits_{x \to a} g(x) = \infty$

3. Evaluating Limits

Limits arise naturally dealing with moving points.

The solutions often involve writing a function f such that the moving point has coordinates $(x, f(x))$. We want to find what happens to the point $(x, f(x))$ when x approaches some given point, i.e., we want to find $\lim\limits_{x \to a} f(x)$.

(1) Substitution Method

If $f(x)$ is an algebraic expression and if $f(a)$ is meaningful, then $\lim\limits_{x \to a} f(x) = f(a)$.

For example, $\lim\limits_{x \to 2} (x^2 - x + 2) = 2^2 - 2 + 2 = 4$ (substituting $x = 2$ into the expression)

However, the expression $f(a)$ will often reduce to something meaningless, such as $\dfrac{0}{0}, \dfrac{\infty}{\infty}, \dfrac{0}{\infty}, \infty \cdot c$

(2) Factoring Method

In case of the form $\dfrac{0}{0}$;

Factor the expression and eliminate the same factor to reduce. Now, use the substitution method into the simplified expression.

For example, $\lim\limits_{x \to 2} \dfrac{x^2-4}{x-2} = \lim\limits_{x \to 2} \dfrac{(x-2)(x+2)}{x-2} = \lim\limits_{x \to 2} (x + 2) = 2 + 2 = 4$

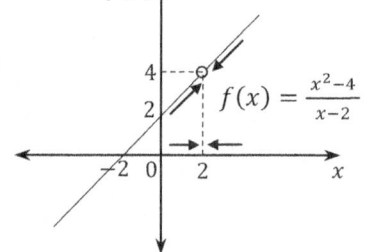

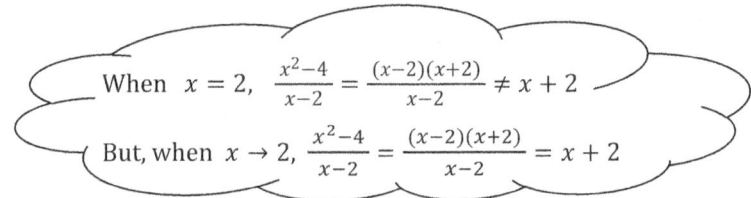

When $x = 2$, $\dfrac{x^2-4}{x-2} = \dfrac{(x-2)(x+2)}{x-2} \neq x + 2$

But, when $x \to 2$, $\dfrac{x^2-4}{x-2} = \dfrac{(x-2)(x+2)}{x-2} = x + 2$

$f(x) = \dfrac{x^2-4}{x-2}$ is undefined at the value $x = 2$; i.e, $f(2)$ does not exist, but $\tilde{f}(x) = x + 2$ is not.

The functions $f(x)$ and $\tilde{f}(x)$ are equivalent except at $x = 2$.

Example   Find $\lim\limits_{x \to 1} \dfrac{x^3 - x^2 - 4x + 4}{x^2 - 1}$

Let $f(x) = \dfrac{x^3 - x^2 - 4x + 4}{x^2 - 1}$.   Then, $f(1)$ is not defined since division by zero is meaningless.

However, if we factor the numerator and denominator of $f(x)$ and reduce the fraction, then we find

that for all x in the domain of f we have $f(x) = \dfrac{x^2(x-1)-4(x-1)}{(x-1)(x+1)} = \dfrac{x^2-4}{x+1} = \dfrac{(x-2)(x+2)}{x+1}$ if $x \neq 1$.

$\therefore$ The expression $f(x) = \dfrac{(x-2)(x+2)}{x+1}$ takes the value $\dfrac{(-1)(3)}{2} = -\dfrac{3}{2}$ at $x = 1$.

Therefore, $f(x) \to -\dfrac{3}{2}$ as $x \to 1$ That is, $\lim\limits_{x \to 1} \dfrac{x^3 - x^2 - 4x + 4}{x^2 - 1} = -\dfrac{3}{2}$

(3) Conjugate Method

In case of the form $\dfrac{0}{0}$ in radical expression, multiply the numerator and denominator by the conjugate of the numerator or denominator.

For example,

1) Evaluate $\lim\limits_{x \to 9} \dfrac{x-9}{\sqrt{x}-3}$.

Substituting $x = 9$, we have $\dfrac{9-9}{\sqrt{9}-3} = \dfrac{0}{0}$ (an indeterminate value)

To apply the conjugate method, multiply the numerator and denominator by the conjugate of the denominator and simplify.

$$\lim\limits_{x \to 9} \dfrac{x-9}{\sqrt{x}-3} = \lim\limits_{x \to 9} \dfrac{(x-9)(\sqrt{x}+3)}{(\sqrt{x}-3)(\sqrt{x}+3)} = \lim\limits_{x \to 9} \dfrac{(x-9)(\sqrt{x}+3)}{x-9} = \lim\limits_{x \to 9} \left(\sqrt{x}+3\right) = \sqrt{9}+3 = 6$$

2) Evaluate $\lim\limits_{x \to 0} \dfrac{\sqrt{2+x}-\sqrt{2}}{\sqrt{2}\,x}$.

Substituting $x = 0$, we have $\dfrac{\sqrt{2+0}-\sqrt{2}}{\sqrt{2}\cdot 0} = \dfrac{0}{0}$ (an indeterminate value)

To apply the conjugate method, multiply the numerator and denominator by the conjugate of the numerator and simplify.

$$\lim\limits_{x \to 0} \dfrac{\sqrt{2+x}-\sqrt{2}}{\sqrt{2}\,x} = \lim\limits_{x \to 0} \dfrac{(\sqrt{2+x}-\sqrt{2})(\sqrt{2+x}+\sqrt{2})}{\sqrt{2}\,x(\sqrt{2+x}+\sqrt{2})} = \lim\limits_{x \to 0} \dfrac{(2+x)-2}{\sqrt{2}\,x(\sqrt{2+x}+\sqrt{2})} = \lim\limits_{x \to 0} \dfrac{x}{\sqrt{2}\,x(\sqrt{2+x}+\sqrt{2})}$$

$$= \lim\limits_{x \to 0} \dfrac{1}{\sqrt{2}(\sqrt{2+x}+\sqrt{2})} = \dfrac{1}{\sqrt{2}(\sqrt{2+0}+\sqrt{2})} = \dfrac{1}{4}$$

(4) More Method

In case of the form $\dfrac{\infty}{\infty}$;

Divide the numerator and denominator by the highest degree term of the denominator.

When the degree of the denominator is greater than the degree of the numerator, the limit at infinity equals 0.

$$\lim\limits_{x \to \infty} \dfrac{x^m}{x^n} = 0 \quad \text{if } m < n$$

$$\lim\limits_{x \to \infty} \dfrac{x^m}{x^n} = \underline{\text{Constant}} \quad \text{if } m = n$$
$$\quad\quad\quad\quad\quad \uparrow\!\text{—— the ratio of the leading coefficients for the numerator and denominator}$$

$$\lim\limits_{x \to \infty} \dfrac{x^m}{x^n} = \infty \quad \text{or} \quad \lim\limits_{x \to \infty} \dfrac{x^m}{x^n} = -\infty \quad \text{if } m > n$$

Example

1) $\displaystyle\lim_{x\to\infty}\frac{2x+1}{2x^2-3x+1}$

When $x \to \infty$, (The denominator) $\to \infty$ and (The numerator) $\to \infty$

Divide both numerator and denominator by the highest degree of the denominator, x^2.

Then, we have $\displaystyle\lim_{x\to\infty}\frac{2x+1}{2x^2-3x+1} = \lim_{x\to\infty}\frac{\frac{2}{x}+\frac{1}{x^2}}{2-\frac{3}{x}+\frac{1}{x^2}} = \frac{0}{2} = 0$

2) $\displaystyle\lim_{x\to\infty}\frac{4x^2-x+1}{3x^2+2x-5}$

Divide both numerator and denominator by the highest degree of the denominator, x^2.

Then, we have $\displaystyle\lim_{x\to\infty}\frac{4x^2-x+1}{3x^2+2x-5} = \lim_{x\to\infty}\frac{4-\frac{1}{x}+\frac{1}{x^2}}{3+\frac{2}{x}-\frac{5}{x^2}} = \frac{4}{3}$

Note:

Since the highest degrees of numerator and denominator are the same,

the limit is the ratio of the leading coefficients. That is, $\displaystyle\lim_{x\to\infty}\frac{4x^2-x+1}{3x^2+2x-5} = \frac{4}{3}$

3) $\displaystyle\lim_{x\to\infty}\frac{4x}{\sqrt{x^2+2}-3} = \lim_{x\to\infty}\frac{\frac{4x}{x}}{\sqrt{\frac{x^2+2}{x^2}}-\frac{3}{x}} = \lim_{x\to\infty}\frac{4}{\sqrt{1+\frac{2}{x^2}}-\frac{3}{x}} = \frac{4}{\sqrt{1+0}-0} = 4$

4) $\displaystyle\lim_{x\to-\infty}\frac{4x}{\sqrt{x^2+2}-3} = \lim_{x\to-\infty}\frac{\frac{4x}{x}}{\left(-\sqrt{\frac{x^2+2}{x^2}}\right)-\frac{3}{x}} = \lim_{x\to-\infty}\frac{4}{\left(-\sqrt{1+\frac{2}{x^2}}\right)-\frac{3}{x}} = \frac{4}{(-\sqrt{1+0})-0} = -4$

$\sqrt{a^2} = \begin{cases} a, & a \geq 0 \\ -a, & a < 0 \end{cases}$

$\therefore$ When $a < 0$, $a = -\sqrt{a^2}$

5) $\displaystyle\lim_{x\to\infty}\frac{2x^2-5x+3}{3x+1}$

Divide both numerator and denominator by the highest degree of the denominator, x.

Then, we have $\displaystyle\lim_{x\to\infty}\frac{2x^2-5x+3}{3x+1} = \lim_{x\to\infty}\frac{2x-5+\frac{3}{x}}{3+\frac{1}{x}} = \infty$

2-2 Special Limits

1. Limits of Trigonometric Functions

For trigonometric functions $y = \sin x$, $y = \cos x$, and $y = \tan x$,

(1) $\displaystyle\lim_{x\to a}\sin x = \sin a$; $\displaystyle\lim_{x\to a}\cos x = \cos a$; $\displaystyle\lim_{x\to a}\tan x = \tan a$

(2) If x approaches 0 and the value inside the sine function matches the denominator,

then $\boxed{\lim_{x \to 0} \dfrac{\sin x}{x} = 1}$

i) When $0 < x < \dfrac{\pi}{2}$

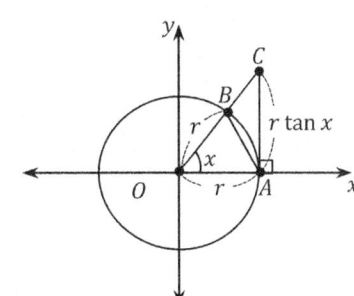

$\sin x = \dfrac{\overline{AC}}{\overline{OC}}$

$\cos x = \dfrac{\overline{OA}}{\overline{OC}} = \dfrac{r}{\overline{OC}}$

$\tan x = \dfrac{\sin x}{\cos x} = \dfrac{\overline{AC}}{r}$; $\overline{AC} = r \tan x$

Note that: The area of ΔOAB < The area of sector OAB < The area of ΔOAC

$\therefore \dfrac{1}{2} r^2 \sin x < \dfrac{1}{2} r^2 x < \dfrac{1}{2} r^2 \tan x$

$\therefore \sin x < x < \tan x$

$\therefore \dfrac{\sin x}{\sin x} < \dfrac{x}{\sin x} < \dfrac{\tan x}{\sin x}$; $1 < \dfrac{x}{\sin x} < \dfrac{1}{\cos x}$; $1 > \dfrac{\sin x}{x} > \cos x$

Since $\cos x \to 1$ as $x \to 0$, $\lim_{x \to 0^+} \cos x = 1$

$\therefore \lim_{x \to 0} \dfrac{\sin x}{x} = 1$

ii) When $-\dfrac{\pi}{2} < x < 0$

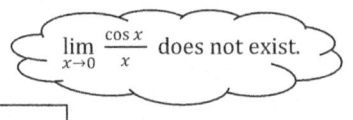

Let $x = -\theta$.   Then, $\theta \to 0^+$ as $x \to 0^-$

$\therefore \lim_{x \to 0^-} \dfrac{\sin x}{x} = \lim_{\theta \to 0^+} \dfrac{\sin(-\theta)}{-\theta} = \lim_{\theta \to 0^+} \dfrac{-\sin\theta}{-\theta} = \lim_{\theta \to 0^+} \dfrac{\sin\theta}{\theta} = 1$ by i)

Therefore, by i) and ii), $\lim_{x \to 0} \dfrac{\sin x}{x} = 1$

$\lim_{x \to 0} \dfrac{\cos x}{x}$ does not exist.

Note that   ① $\boxed{\lim_{x \to 0} \dfrac{\tan x}{x} = 1 \; ; \; \lim_{x \to 0} \dfrac{x}{\tan x} = 1}$ ② $\boxed{\lim_{x \to 0} \dfrac{\cos x - 1}{x} = 0}$

① $\lim_{x \to 0} \dfrac{\tan x}{x} = \lim_{x \to 0} \left(\dfrac{1}{\cos x} \cdot \dfrac{\sin x}{x} \right) = 1 \cdot 1 = 1$

② Since $\dfrac{\cos x - 1}{x} = \dfrac{\cos x - 1}{x} \cdot \dfrac{\cos x + 1}{\cos x + 1} = \dfrac{\cos^2 x - 1}{x(\cos x + 1)} = \dfrac{-\sin^2 x}{x(\cos x + 1)} = \dfrac{\sin x}{x} \cdot \left(\dfrac{-\sin x}{\cos x + 1} \right)$,

$\lim_{x \to 0} \dfrac{\cos x - 1}{x} = \lim_{x \to 0} \left\{ \dfrac{\sin x}{x} \cdot \left(\dfrac{-\sin x}{\cos x + 1} \right) \right\} = \lim_{x \to 0} \dfrac{\sin x}{x} \cdot - \left(\lim_{x \to 0} \dfrac{\sin x}{\cos x + 1} \right) = 1 \cdot \left(-\dfrac{0}{2} \right) = 0$

(3) $\boxed{\begin{array}{l} \lim_{x \to 0} \dfrac{\sin bx}{ax} = \dfrac{b}{a} \; ; \; \lim_{x \to 0} \dfrac{\sin bx}{\sin ax} = \dfrac{b}{a} \\[2mm] \lim_{x \to 0} \dfrac{\tan bx}{ax} = \dfrac{b}{a} \; ; \; \lim_{x \to 0} \dfrac{\tan bx}{\tan ax} = \dfrac{b}{a} \\[2mm] \lim_{x \to 0} \dfrac{\sin x}{\tan x} = 1 \; ; \; \lim_{x \to 0} \dfrac{\sin bx}{\tan ax} = \dfrac{b}{a} \end{array}}$

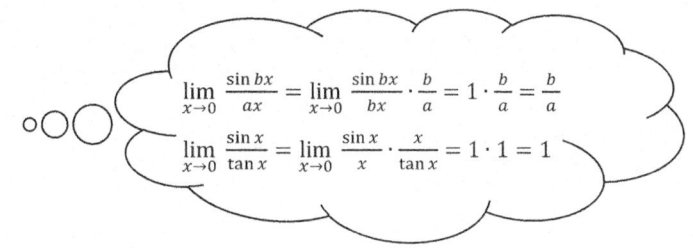

$\lim_{x \to 0} \dfrac{\sin bx}{ax} = \lim_{x \to 0} \dfrac{\sin bx}{bx} \cdot \dfrac{b}{a} = 1 \cdot \dfrac{b}{a} = \dfrac{b}{a}$

$\lim_{x \to 0} \dfrac{\sin x}{\tan x} = \lim_{x \to 0} \dfrac{\sin x}{x} \cdot \dfrac{x}{\tan x} = 1 \cdot 1 = 1$

Example

1) $\displaystyle\lim_{x \to 0} \frac{\sin 2x}{\sin x}$

If $x \to a$ $(a \neq 0)$, then let $x - a = t$ so that $t \to 0$.

$\because$ Since $\sin 2x = 2 \sin x \cos x$,

$$\lim_{x \to 0} \frac{\sin 2x}{\sin x} = \lim_{x \to 0} \frac{2 \sin x \cos x}{\sin x} = \lim_{x \to 0} 2 \cos x = 2\lim_{x \to 0} \cos x = 2 \cdot 1 = 2$$

2) $\displaystyle\lim_{x \to 0} \frac{\cos 3x - \cos x}{\sin 2x}$

$\because$ Since $\cos 3x = 4 \cos^3 x - 3 \cos x$,

$$\lim_{x \to 0} \frac{\cos 3x - \cos x}{\sin 2x} = \lim_{x \to 0} \frac{4 \cos^3 x - 3 \cos x - \cos x}{2 \sin x \cos x} = \lim_{x \to 0} \frac{4 \cos^3 x - 4 \cos x}{2 \sin x \cos x} = \lim_{x \to 0} \frac{2(\cos^2 x - 1)}{\sin x}$$

$$= \lim_{x \to 0} \frac{2(-\sin^2 x)}{\sin x} = \lim_{x \to 0} (-2\sin x) = 0$$

Alternative approach:

Since $\cos A - \cos B = -2 \sin\frac{A+B}{2} \sin\frac{A-B}{2}$,

$$\cos 3x - \cos x = -2 \sin\frac{3x+x}{2} \sin\frac{3x-x}{2} = -2 \sin 2x \sin x$$

$$\therefore \lim_{x \to 0} \frac{\cos 3x - \cos x}{\sin 2x} = \lim_{x \to 0} \left(\frac{-2\sin 2x \sin x}{\sin 2x}\right) = \lim_{x \to 0} (-2 \sin x) = 0$$

3) $\displaystyle\lim_{x \to 0} x \sin\frac{1}{x}$

$\because$ Since $\left|\sin\frac{1}{x}\right| \leq 1$, $0 \leq \left|x \sin\frac{1}{x}\right| = |x|\left|\sin\frac{1}{x}\right| \leq |x|$

Since $\displaystyle\lim_{x \to 0} |x| = 0$, $\displaystyle\lim_{x \to 0} \left|x \sin\frac{1}{x}\right| = 0$

$\therefore \displaystyle\lim_{x \to 0} x \sin\frac{1}{x} = 0$

4) $\displaystyle\lim_{x \to 0} \sin x \cos\frac{1}{x}$

$\because$ Since $\left|\cos\frac{1}{x}\right| \leq 1$, $0 \leq \left|\sin x \cos\frac{1}{x}\right| = |\sin x|\left|\cos\frac{1}{x}\right| \leq |\sin x|$

Since $\displaystyle\lim_{x \to 0} |\sin x| = 0$, $\displaystyle\lim_{x \to 0} \left|\sin x \cos\frac{1}{x}\right| = 0$

$\therefore \displaystyle\lim_{x \to 0} \sin x \cos\frac{1}{x} = 0$

5) $\displaystyle\lim_{x \to 0} \frac{\sin 3x}{4x}$

$\because \displaystyle\lim_{x \to 0} \frac{\sin 3x}{4x} = \lim_{x \to 0} \frac{\sin 3x}{3x} \cdot \frac{3x}{4x} = 1 \cdot \frac{3}{4} = \frac{3}{4}$

6) $\displaystyle\lim_{x \to 0} \frac{\tan x}{x}$

$\because \displaystyle\lim_{x \to 0} \frac{\tan x}{x} = \lim_{x \to 0} \frac{\sin x}{x \cos x} = \lim_{x \to 0} \frac{\sin x}{x} \cdot \frac{1}{\cos x} = \lim_{x \to 0} \frac{\sin x}{x} \cdot \lim_{x \to 0} \frac{1}{\cos x} = 1 \cdot 1 = 1$

2. Limits of Exponential Functions

Consider an exponential function $y = a^x$ $(a > 0, a \neq 1)$

When $a > 1$, the graph of the function rises as x increases, called *monotone increasing*.

When $0 < a < 1$, the graph of the function falls as x increases, called *monotone decreasing*.

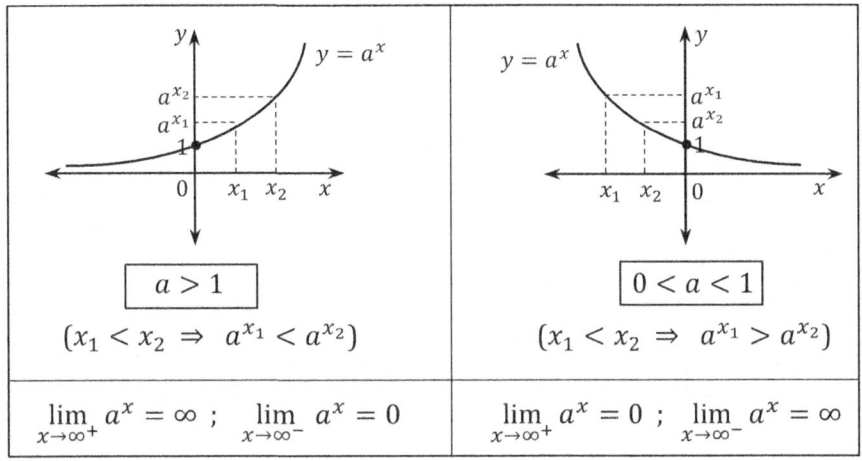

$a > 1$	$0 < a < 1$
$(x_1 < x_2 \Rightarrow a^{x_1} < a^{x_2})$	$(x_1 < x_2 \Rightarrow a^{x_1} > a^{x_2})$
$\lim\limits_{x \to \infty^+} a^x = \infty$; $\lim\limits_{x \to \infty^-} a^x = 0$	$\lim\limits_{x \to \infty^+} a^x = 0$; $\lim\limits_{x \to \infty^-} a^x = \infty$

Example Find the limit:

(1) $\lim\limits_{x \to \infty} \dfrac{\sqrt{3^x}}{2^x} = \lim\limits_{x \to \infty} \dfrac{(\sqrt{3})^x}{2^x} = \lim\limits_{x \to \infty} \left(\dfrac{\sqrt{3}}{2}\right)^x = 0 \quad \left(\because 0 < \dfrac{\sqrt{3}}{2} < 1\right)$

(2) $\lim\limits_{x \to \infty} \dfrac{3^x}{2^x} = \lim\limits_{x \to \infty} \left(\dfrac{3}{2}\right)^x = \infty \quad \left(\because \dfrac{3}{2} > 1\right)$

(3) $\lim\limits_{x \to \infty^-} \left(\dfrac{2}{3}\right)^x$

Let $x = -t$.

Since $x \to \infty^-$, $t \to \infty^+$

$\therefore \lim\limits_{x \to \infty^-} \left(\dfrac{2}{3}\right)^x = \lim\limits_{t \to \infty^+} \left(\dfrac{2}{3}\right)^{-t} = \lim\limits_{t \to \infty} \left(\dfrac{3}{2}\right)^t = \infty \quad \left(\because \dfrac{3}{2} > 1\right)$

3. Limits of Logarithmic Functions

By the definition of a logarithm, the logarithmic function $f(x) = \log_a x$ is the inverse of the exponential function $g(x) = a^x$.

That is, $f(g(x)) = \log_a a^x = x$ and $g(f(x)) = a^{\log_a x} = x$.

By reflecting the graph of $y = a^x$, we obtain the graph of $y = \log_a x$.

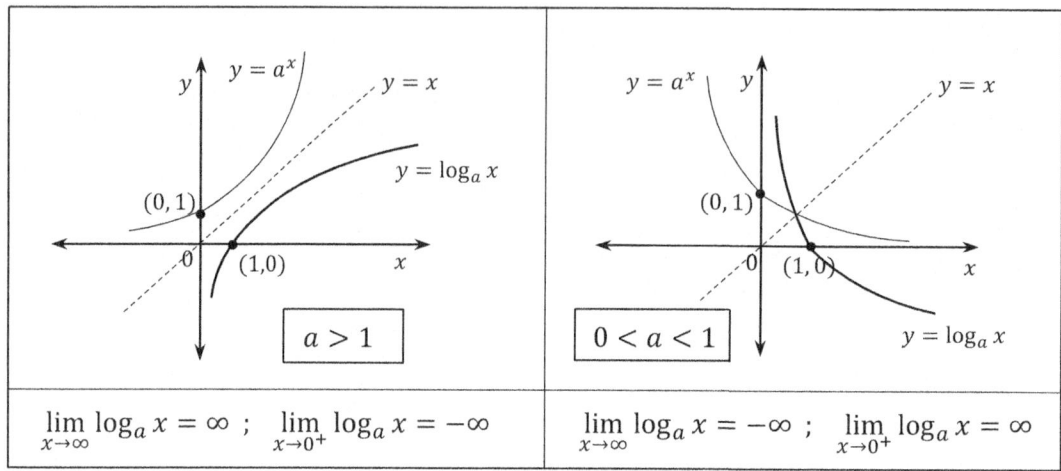

$\displaystyle\lim_{x \to \infty} \log_a x = \infty$; $\displaystyle\lim_{x \to 0^+} \log_a x = -\infty$	$\displaystyle\lim_{x \to \infty} \log_a x = -\infty$; $\displaystyle\lim_{x \to 0^+} \log_a x = \infty$

Example   Find the limit:

(1) $\displaystyle\lim_{x \to 1} \log_2 |x - 1| = -\infty$

(2) $\displaystyle\lim_{x \to 2} \log_{\frac{1}{2}} |x - 2| = \infty$

4. Irrational Number e and the Natural Logarithm

For an expression $\left(1 + \dfrac{1}{n}\right)^n$, $\left(1 + \dfrac{1}{n}\right)^n$ gets closer and closer to $2.718281\cdots$ as n gets larger and larger.

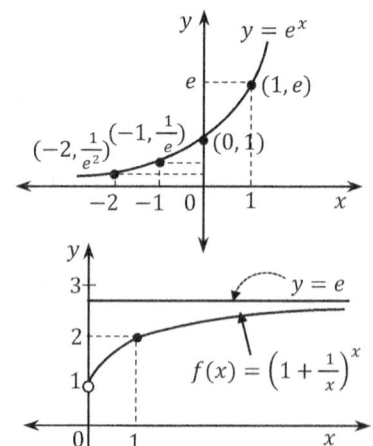

Given an exponential function $y = a^x$,

a particular number a ($a > 1$) which is an irrational number and approaches $e \approx 2.718281$ is called the *natural base*.

The function $f(x) = e^x$ is the *natural exponential function*.

$$\left(1 + \frac{1}{x}\right)^x \to e \ \text{ as } \ x \to \infty$$

As x increases,

the graph of f gets closer and closer to the line $y = e$.

Consider the limit of the function $y = (1 + x)^{\frac{1}{x}}$ as x approaches 0.

(1) Definition

$$e = \lim_{x \to 0}(1 + x)^{\frac{1}{x}} = \lim_{x \to \infty}\left(1 + \frac{1}{x}\right)^x = 2.718281\cdots$$

Note:

For $\displaystyle\lim_{x \to \infty}\left(1 + \frac{1}{x}\right)^x$, let $\dfrac{1}{x} = t$. Then, $x = \dfrac{1}{t}$

Since $t \to 0$ as $x \to \infty$, $\displaystyle\lim_{x \to \infty}\left(1 + \frac{1}{x}\right)^x = \lim_{t \to 0}(1 + t)^{\frac{1}{t}} = e$

(2) If the base is e, then the logarithmic function $\log_e x$ is called the *natural logarithm* of x.

We write: $\log_e x = \ln x$.

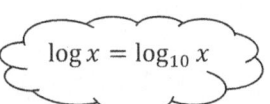

1) $\displaystyle\lim_{x \to 0}\frac{\ln(1+x)}{x} = \lim_{x \to 0}\ln(1 + x)^{\frac{1}{x}} = \ln e = 1$

2) $\displaystyle\lim_{x \to 0}\frac{e^x - 1}{x} = 1$

3) $\displaystyle\lim_{x \to 0}\frac{\log_a(1+x)}{x} = \lim_{x \to 0}\log_a(1 + x)^{\frac{1}{x}} = \log_a e = \frac{1}{\log_e a} = \frac{1}{\ln a}$

4) $\displaystyle\lim_{x \to 0}\frac{a^x - 1}{x} = \ln a$

$\log x = \log_{10} x$

Example   ① $\ln e^3 = \log_e e^3 = 3\log_e e = 3$

② $\ln\sqrt{e} = \log_e e^{\frac{1}{2}} = \frac{1}{2}\log_e e = \frac{1}{2}$

③ $\ln\dfrac{1}{\sqrt[3]{e}} = \log_e e^{-\frac{1}{3}} = -\frac{1}{3}\log_e e = -\frac{1}{3}$

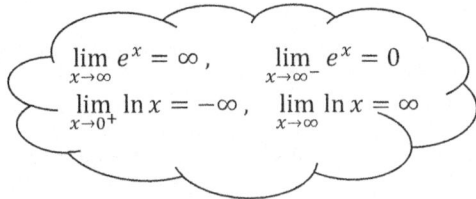

Example   Find the value of x for the following:

(1) $\ln x = 3$

∵ If $\ln x = 3$, then $\log_e x = 3$ ∴ $x = e^3$

(2) $\ln(x + 2) = 1$

∵ If $\ln(x + 2) = 1$, then $\log_e(x + 2) = 1$ ∴ $x + 2 = e^1 = e$ ∴ $x = e - 2$

(3) $e^x = 2$

∵ If $e^x = 2$, then, $x = \log_e 2$

Example Find the following limits:

(1) $\displaystyle\lim_{x\to 0}(1+2x)^{\frac{1}{x}} = \lim_{x\to 0}\left\{(1+2x)^{\frac{1}{2x}}\right\}^2 = e^2$

(2) $\displaystyle\lim_{x\to 0}(1+x)^{\frac{3}{x}} = \lim_{x\to 0}\left\{(1+x)^{\frac{1}{x}}\right\}^3 = e^3$

(3) $\displaystyle\lim_{x\to 0}(1+2x)^{\frac{1}{3x}} = \lim_{x\to 0}\left\{(1+2x)^{\frac{1}{2x}}\right\}^{\frac{2}{3}} = e^{\frac{2}{3}}$

(4) $\displaystyle\lim_{x\to\infty}\left(1+\frac{3}{x}\right)^{x} = \lim_{x\to\infty}\left\{\left(1+\frac{3}{x}\right)^{\frac{x}{3}}\right\}^3 = e^3$

(5) $\displaystyle\lim_{x\to\infty}\left(1+\frac{1}{x}\right)^{2x} = \lim_{x\to\infty}\left\{\left(1+\frac{1}{x}\right)^{x}\right\}^2 = e^2$

(6) $\displaystyle\lim_{x\to\infty}\left(1+\frac{2}{x}\right)^{3x} = \lim_{x\to\infty}\left\{\left(1+\frac{2}{x}\right)^{\frac{x}{2}}\right\}^6 = e^6$

2-3 Continuity

1. Continuous Functions

(1) Open and Closed Intervals

Bounded and Unbounded Intervals

For any real numbers a and b such that $a < b$, bounded and unbounded intervals on the real number line are as follows (where a and b are the end points of each interval).

$a \leq x \leq b \quad\Rightarrow\quad [a,b]$: Closed interval

$a < x < b \quad\Rightarrow\quad (a,b)$: Open interval

$a \leq x < b \quad\Rightarrow\quad [a,b)$: Half-Open interval

$a < x \leq b \quad\Rightarrow\quad (a,b]$: Half-Open interval

— Bounded Intervals

$x \geq a \qquad\Rightarrow\quad [a,\infty)$: Half-Open interval

$x > a \qquad\Rightarrow\quad (a,\infty)$: Open interval

$x \leq b \qquad\Rightarrow\quad (-\infty,b]$: Half-Open interval

$x < b \qquad\Rightarrow\quad (-\infty,b)$: Open interval

— Unbounded Intervals

$(-\infty,\infty)$: Whole real line

Example Find the domain of each function.

1) $f(x) = x^2$ The domain of the function is $(-\infty,\infty)$.

2) $f(x) = \dfrac{x^2-1}{x-1}$

   Since the function is undefined at $x = 1$, the domain of the function is $(-\infty,1) \cup (1,\infty)$.

3) $f(x) = \frac{1}{x^2-1}$

Since the function is undefined at $x^2 - 1 = 0$; $(x+1)(x-1) = 0$,

the domain of the function is $(-\infty, -1) \cup (-1, 1) \cup (1, \infty)$.

4) $f(x) = \sqrt{x-2}$

Since $x - 2 \geq 0$, the domain of the function is $[2, \infty)$.

(2) Continuous Functions

A function f, defined on an open interval containing a, is *continuous* at a point a if $\lim\limits_{x \to a} f(x) = f(a)$.

Note: $f(x)$ is said to be continuous at a if ① $\lim\limits_{x \to a} f(x)$ exists

② $f(a)$ exists, and

③ $\lim\limits_{x \to a} f(x) = f(a)$.

Continuity of a function in an interval implies that the graph of the function in the interval is an uninterrupted curve.

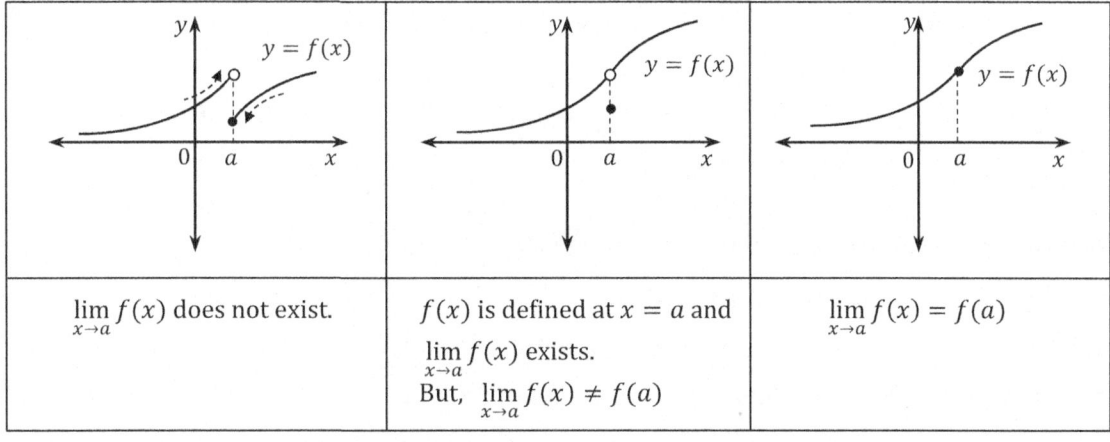

$\lim\limits_{x \to a} f(x)$ does not exist.	$f(x)$ is defined at $x = a$ and $\lim\limits_{x \to a} f(x)$ exists. But, $\lim\limits_{x \to a} f(x) \neq f(a)$	$\lim\limits_{x \to a} f(x) = f(a)$

The sine and cosine functions are continuous at zero.

∵ Since $\lim\limits_{x \to 0} \sin x = 0 = \sin 0$ and $\lim\limits_{x \to 0} \cos x = 1 = \cos 0$, the condition for continuity is satisfied.

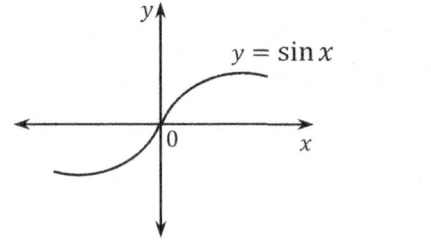 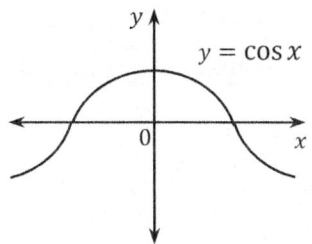

1) Consider a function $f(x) = \frac{\sin x}{x}$.

This function is defined when x is any real number except 0.

Note that $\lim\limits_{x \to 0} \frac{\sin x}{x} = 1$

But, $\frac{\sin 0}{0}$ is nonsense.

That means we can find the limit as $x \to 0$, but the function is not defined at $x = 0$.

If we wished, we could have a new function $\tilde{f}$, where $\tilde{f}(x) = \begin{cases} \frac{\sin x}{x}, & x \neq 0 \\ 4\pi, & x = 0 \end{cases}$

Then, the new function $\tilde{f}$ is defined everywhere.

Since the function $f(x) = \frac{\sin x}{x}$ ($x \neq 0$) is not defined at the origin, the function f is not continuous

at $x = 0$ and hence must be discontinuous there.

The new function $\tilde{f}$ is discontinuous at $x = 0$, since $\lim\limits_{x \to 0} \tilde{f}(x) = 1 \neq \tilde{f}(0) = 4\pi$.

The function g defined by $g(x) = \begin{cases} \frac{\sin x}{x}, & x \neq 0 \\ 1, & x = 0 \end{cases}$ is continuous at $x = 0$, since $\lim\limits_{x \to 0} g(x) = 1 = g(0)$.

For the function g, which is continuous at $x = 0$, we can evaluate $\lim\limits_{x \to 0} g(x)$ by substituting $x = 0$.

2) Consider a function $h(x) = |x|$.

Since $h(x) = \begin{cases} x, & x \geq 0 \\ -x, & x < 0 \end{cases}$, the function h is continuous at all real numbers different from 0.

But, $\lim\limits_{x \to 0} |x| = 0 = |0|$.

Therefore, $|x|$ is also continuous at 0 and hence h is continuous everywhere.

Note:

① A polynomial function is continuous at every real number a.

② A rational function is continuous at every real number a in its domain; i.e., except where its

denominator is zero.

Example

(1) Prove that the polynomial function $f(x) = x$ is continuous for all x.

　　∵ For any real number a, $\lim\limits_{x \to a} f(x) = \lim\limits_{x \to a} x = a$ and $f(a) = a$

　　∴ $f(x)$ is continuous at $x = a$.

　　Since a is any real number, $f(x)$ is continuous for all x.

　　(Similarly, the constant function $f(x) = c$ is continuous for all x.)

(2) How should $f(x) = \dfrac{x^2-4}{x-2}$ $(x \neq 2)$ be defined at $x = 2$ in order to make it continuous there?

$\because \lim\limits_{x \to 2} \dfrac{x^2-4}{x-2} = \lim\limits_{x \to 2} \dfrac{(x-2)(x+2)}{x-2} = \lim\limits_{x \to 2}(x+2) = 4$

Therefore, we define $f(2) = 4$.

That is, $f(x) = \begin{cases} \dfrac{x^2-4}{x-2}, & x \neq 2 \\ 4, & x = 2 \end{cases}$

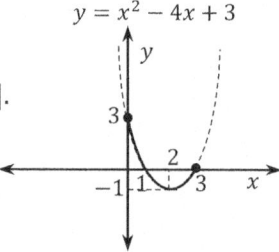

$y = x^2 - 4x + 3$

Note:

Consider a function $f(x) = x^2 - 4x + 3$, defined on a closed interval $[0, 3]$.

① $f(x)$ is continuous for all x in the open interval $(0, 3)$.

② $\lim\limits_{x \to 0^+} f(x) = 3 = f(0)$ and $\lim\limits_{x \to 3^-} f(x) = 0 = f(3)$

Therefore, $f(x)$ is continuous for all x in the closed interval $[0, 3]$.

(3) Discontinuous at $x = a$

If a function f fails to be continuous at a point $x = a$ of its domain, it is said to be *discontinuous* there. The point a is called a *point of discontinuity* or, simply a *discontinuity* of f.

Note:

If ① $f(x)$ is not defined at $x = a$ or

② $\lim\limits_{x \to a} f(x)$ does not exist or

③ $f(x)$ is defined at $x = a$, $\lim\limits_{x \to a} f(x)$ exists, but $\lim\limits_{x \to a} f(x) \neq f(a)$,

then $f(x)$ is not continuous at $x = a$.

Example Examine for the discontinuity of the function.

(1) $f(x) = \dfrac{1}{x-1}$

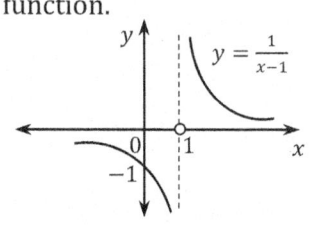

$y = \dfrac{1}{x-1}$

  $f(1)$ does not exist.

  $\therefore$ $f(x)$ is discontinuous at $x = 1$.

(2) $f(x) = \dfrac{x^2-1}{x-1}$

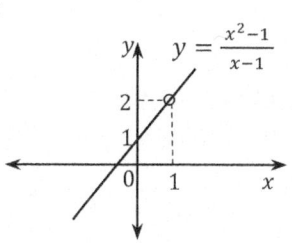

$y = \dfrac{x^2-1}{x-1}$

  $f(x) = \dfrac{x^2-1}{x-1} = \dfrac{(x-1)(x+1)}{x-1}$

  If $x \neq 1$, then $f(x) = x + 1$

  If $x = 1$, then $f(1)$ does not exist. $\therefore$ $f(x)$ is discontinuous at $x = 1$.

(3) $f(x) = [x]$ ($[x]$ is the greatest integer in x)

$f(x) = -1$ in the interval $[-1, 0)$

$f(x) = 0$ in the interval $[0, 1)$

$f(x) = 1$ in the interval $[1, 2)$

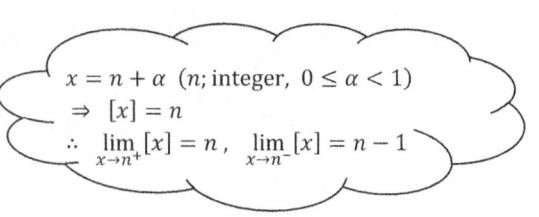

$$x = n + \alpha \ (n; \text{integer}, \ 0 \le \alpha < 1)$$
$$\Rightarrow [x] = n$$
$$\therefore \lim_{x \to n^+} [x] = n, \quad \lim_{x \to n^-} [x] = n - 1$$

$\vdots$

Note that $f(1) = 1$. But, $\lim_{x \to 1^-} f(x) = 0$, $\lim_{x \to 1^+} f(x) = 1$

$\therefore \ \lim_{x \to 1} f(x)$ does not exist.

$\therefore \ f(x)$ is discontinuous at $x = 1$.

Similarly, $f(x)$ is discontinuous at $x = \cdots, -1, 0, 2, 3, \cdots$

Therefore, $f(x)$ is discontinuous at all integers.

(4) $f(x) = \begin{cases} x^2 + 1, & x \neq 0 \\ 0, & x = 0 \end{cases}$

Since $f(0) = 0$, $f(x)$ is defined at $x = 0$.

Since $\lim_{x \to 0} f(x) = 1$, $\lim_{x \to 0} f(x)$ exists.

But, $\lim_{x \to 0} f(x) \neq f(0)$

$\therefore \ f(x)$ is discontinuous at $x = 0$.

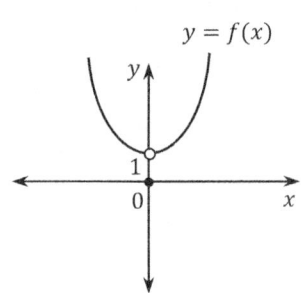

2. Properties of Continuous Functions

(1) If the functions f and g are continuous at $x = a$, then the functions

① $kf(x)$, k: Constant

② $f(x) \pm g(x)$

③ $f(x)g(x)$

④ $|f(x)|$

$$\lim_{x \to a}(f \pm g)(x) = \lim_{x \to a} f(x) \pm \lim_{x \to a} g(x) = f(a) \pm g(a) = (f \pm g)(a)$$
$$\lim_{x \to a}(fg)(x) = \lim_{x \to a} f(x) \cdot \lim_{x \to a} g(x) = f(a) \cdot g(a) = (fg)(a)$$

are also continuous at $x = a$.

If $g(a) \neq 0$, then the function

⑤ $\dfrac{f(x)}{g(x)}$ is continuous at $x = a$.

(2) A polynomial function:

$f(x) = a_n x^n + a_{n-1} x^{n-1} + a_{n-2} x^{n-2} + \cdots\cdots + a_1 x + a_0$ defined over the real numbers

is continuous for all values of $x, (-\infty, \infty)$.

A rational function $\dfrac{f(x)}{g(x)}$ defined over the real number is continuous for all real values of x

except for the zeroes of $g(x)$.

Continuity Properties of Other Functions of a Real Variable:

Functions	Domain of Continuity
$\sqrt{f(x)}$ (f; Polynomial function)	All $f(x) > 0$
a^x ($a > 0$)	All x
$\log x$	All $x > 0$
$\|x\|$	All x
$[x]$	All x except x equal to an integer
$\sin x$	All x
$\cos x$	All x
$\tan x$	All x except $x = n\pi + \dfrac{\pi}{2}$ (n; integer)

Note:

Polynomial, rational, exponential, logarithmic, and trigonometric functions are always continuous over their domains.

(3) Composition of continuous functions:

Composite Limit Theorem

If $\displaystyle\lim_{x\to a} g(x) = L$ and if f is continuous at L, then $\displaystyle\lim_{x\to a} f\{g(x)\} = f\{\lim_{x\to a} g(x)\} = f(L)$

If f and g are two continuous functions of x, then the composite function h defined by $h = f(g(x))$ is continuous and we write $\displaystyle\lim_{x\to a} f(g(x)) = f\left(\lim_{x\to a} g(x)\right) = f(g(a))$

> If g is continuous at a and f is continuous at $g(a)$, then the composite function $f \circ g$ is continuous at a.
> $(f \circ g)(x) = f(g(x))$

Example

(1) $\displaystyle\lim_{x\to 0} \sin\left(x + \frac{\pi}{2}\right) = \sin\left[\lim_{x\to 0}\left(x + \frac{\pi}{2}\right)\right] = \sin\left(\frac{\pi}{2}\right) = 1$

(2) $\displaystyle\lim_{x\to 1} \log(x^2 + 3) = \log\left[\lim_{x\to 1}(x^2 + 3)\right] = \log 4$

Example Show that the functions (1) $h(x) = \frac{x^2-x-2}{x^2+1}$ (2) $h(x) = |x^2 - 2x + 4|$ are continuous at each real number.

(1) Since $x^2 + 1 \neq 0$, the rational function $h(x) = \frac{x^2-x-2}{x^2+1}$ is continuous at all real numbers.

(2) Let $f(x) = |x|$ and $g(x) = x^2 - 2x + 4$.

Then, f and g are continuous at each real number.

∴ Their composite $h(x) = f(g(x)) = |x^2 - 2x + 4|$ is also continuous at each real number.

3. The Intermediate-Value Property

The continuity of f at $x = a$ means that if x is near a, then $f(x)$ is near $f(a)$.

Thus, we have the following properties:

(1) If f is continuous in a closed interval I, then $f(x)$ is bounded in the interval I.

(2) If f is defined in an interval I and $x = c$ be a point of I, then

    ① $f(c)$ is a maximum, M, of f in I if and only if $f(c) \geq f(x)$ for all x in I.

    ② Similarly, $f(c)$ is a minimum, m, of f in I if and only if $f(c) \leq f(x)$ for all x in I.

(3) If f is continuous in a closed interval I,

    then there is a point c of I at which $f(c)$ is a maximum of f in I.

(4) If f is continuous in a closed interval I,

    then there is a point c of I at which $f(c)$ is a minimum of f in I.

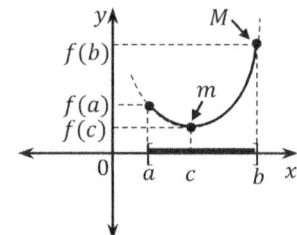

Example

(1) $f(x) = \begin{cases} x^2, & x \neq 0 \\ 1, & x = 0 \end{cases}$

    Domain of the function is $(-\infty, \infty)$

    Point of discontinuity: $x = 0$

    f has neither maximum nor minimum.

(2) $f(x) = \frac{1}{x}$

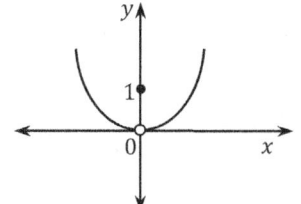

    i) When $x > 0$, domain of the function is $(0, \infty)$.

      f is continuous.

      f has neither maximum nor minimum.

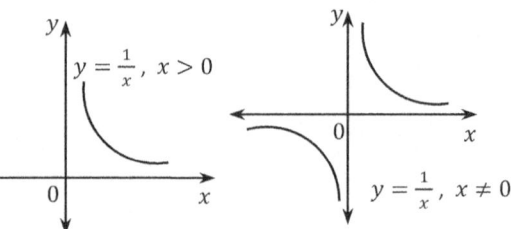

    ii) When $x \neq 0$, domain of the function is $(-\infty, 0) \cup (0, \infty)$.

      Point of discontinuity: $x = 0$

      f has neither maximum nor minimum.

The Intermediate-Value Theorem

> If f is continuous on closed interval $[a, b]$ and $f(a) \neq f(b)$, then
> for each k between $f(a)$ and $f(b)$,
> there is at least one number c between a and b such that $f(c) = k$.

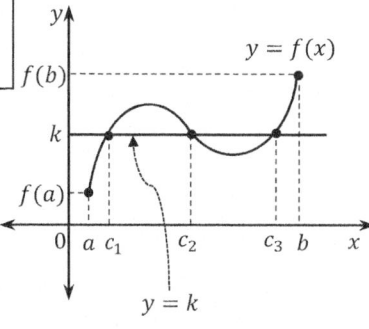

The Fixed point Theorem

> If $f(x)$ is continuous on closed interval $[a, b]$
> such that $f(a)$ and $f(b)$ also belong to the interval $[a, b]$, then
> there exists some value c in the interval $[a, b]$ such that $f(c) = c$.

Note:

If f is continuous on closed interval $[a, b]$ and

$f(a)f(b) < 0$ ($f(a)$ and $f(b)$ have different signs),

then there is at least one number c in the open interval (a, b) such that $f(c) = 0$.

Exercises

#1 Evaluate the following limits:

(1) $\lim\limits_{x \to 2} 3x^2$

(2) $\lim\limits_{x \to 3} x(2x - 1)$

(3) $\lim\limits_{x \to 2} \dfrac{x^2}{x+1}$

(4) $\lim\limits_{x \to 1} \sqrt{x + 3}$

(5) $\lim\limits_{x \to 0} \cos x$

(6) $\lim\limits_{x \to 1} \sin \dfrac{\pi x}{2}$

(7) $\lim\limits_{x \to 0} \dfrac{2^{x+1}}{2^x + 3^x}$

(8) $\lim\limits_{x \to 10} \left(\log x^2 - \log \dfrac{1}{x} \right)$

(9) $\lim\limits_{x \to 0} \dfrac{1}{\sqrt{x^2}}$

(10) $\lim\limits_{x \to 1} \dfrac{x}{(x-1)^2}$

(11) $\lim\limits_{x \to 2} \left(3 - \dfrac{x}{|x-2|} \right)$

(12) $\lim\limits_{x \to 3} \dfrac{x^2 + x - 12}{x - 3}$

(13) $\lim\limits_{x \to 1} \left(\dfrac{5x^2 - 7x + 2}{x - 1} + \dfrac{3 - 2x}{x + 1} \right)$

(14) $\lim\limits_{x \to a} \dfrac{x^3 - ax^2 + a^2 x - a^3}{x - a}$

(15) $\lim\limits_{x \to 1} \dfrac{\sqrt{x+8} - 3}{x - 1}$

(16) $\lim\limits_{x \to 3} \dfrac{x - 3}{\sqrt{x-3}}$

(17) $\lim\limits_{x \to -8} \dfrac{x + 8}{\sqrt[3]{x+2}}$

(18) $\lim\limits_{x \to 1} \dfrac{\sqrt[3]{x} - 1}{x - 1}$

(19) $\lim\limits_{x \to \infty} \dfrac{2x - 3}{3x^2 - 2x + 1}$

(20) $\lim\limits_{x \to \infty} \dfrac{5x^3 + 2x}{2x^3 - 4x^2 + 5x}$

(21) $\lim\limits_{x \to \infty} \dfrac{3x^3 - 2x + 4}{x^2 + 1}$

(22) $\lim\limits_{x \to \infty} \dfrac{\sqrt{x^2+1} - 1}{x + 1}$

(23) $\lim\limits_{x \to -\infty} \dfrac{\sqrt{x^2+1} - 1}{x + 1}$

(24) $\lim\limits_{x \to \infty} (2x^3 - 3x^2 + 4x - 1)$

(25) $\lim\limits_{x \to \infty} (\sqrt{x^2 + 2x + 3} - x)$

(26) $\lim\limits_{x \to 0} \dfrac{1}{x} \left(1 + \dfrac{1}{x-1} \right)$

(27) $\lim\limits_{x \to 0} \dfrac{1}{x} \left(\dfrac{1}{\sqrt{x+1}} - 1 \right)$

(28) $\lim\limits_{x \to -\infty} \dfrac{x + 1}{\sqrt{x^2 + x} - x}$

#2 Find the value of the constant a such that the following expression:

(1) $\lim\limits_{x \to 2} \dfrac{x^2 - 2x + 2}{3x^2 + ax + 1} = 2$

(2) $\lim\limits_{x \to \infty} \dfrac{3x^3 + 4x - 1}{ax^3 + 2x^2 + 3} = \dfrac{1}{2}$

(3) $\lim\limits_{x \to \infty} \dfrac{a \, x^4}{4^{x+1} - 3^x} = 4$

(4) $\lim\limits_{x \to \infty} \dfrac{ax}{\sqrt{x^2+1} - 1} = 2$

(5) $\lim\limits_{x \to \infty} \{ \log(ax + 1) - \log x \} = 1$

#3 Find the values of the constants a and b such that the following expression:

(1) $\lim\limits_{x \to 1} \dfrac{ax - 3x^2}{x - 1} = b$

(2) $\lim\limits_{x \to 1} \dfrac{x^2 + ax + b}{x - 1} = 5$

(3) $\lim\limits_{x \to -3} \dfrac{\sqrt{x^2 - x - 3} + ax}{x + 3} = b$

#4 Find the expression of $f(x)$ such that $\lim\limits_{x \to \infty} \dfrac{f(x)}{2x^2 + x + 1} = 1$ and $\lim\limits_{x \to 2} \dfrac{f(x)}{x^2 - x - 2} = 1$.

#5 Determine whether the limits exist.

(1) $\lim\limits_{x \to 0} \dfrac{1}{x}$

(3) $\lim\limits_{x \to 1} \dfrac{|x^2 - 1|}{|x - 1|}$

(2) $\lim\limits_{x \to \infty} \sin \dfrac{1}{x}$

(4) $\lim\limits_{x \to \infty} (\sqrt{x^2 - x} - x)$

#6 For functions $f(x), g(x),$ and $h(x)$ which are defined on all real numbers, determine if the following statements are true or false.

(1) If $\lim\limits_{x \to 0} f(x) = 1$, then $f(0) = 1$.

(2) If $\lim\limits_{x \to 1} f(x) = 1$, then $\lim\limits_{x \to \infty} f\left(1 + \dfrac{1}{x}\right) = 1$.

(3) If $f(x) < g(x) < h(x)$, $\lim\limits_{x \to 0} f(x) = 0$, and $\lim\limits_{x \to 0} h(x) = 0$, then $\lim\limits_{x \to 0} g(x) = 0$.

(4) If $\lim\limits_{x \to \infty} xf(x)$ exists, then $\lim\limits_{x \to \infty} f(x)$ exists.

(5) If $\lim\limits_{x \to \infty} \dfrac{1}{f(x)}$ exists, then $\lim\limits_{x \to \infty} f(x)$ exists.

(6) If $\lim\limits_{x \to a} f(x) = \infty$ and $\lim\limits_{x \to a} f(x)\,g(x)$ exists, then $\lim\limits_{x \to a} g(x) = 0$.

(7) If $\lim\limits_{x \to a} f(x)$ and $\lim\limits_{x \to a} f(x)\,g(x)$ exist, then $\lim\limits_{x \to a} g(x)$ exists.

(8) If $\lim\limits_{x \to a} g(x)$ and $\lim\limits_{x \to \infty} \dfrac{f(x)}{g(x)}$ exist, then $\lim\limits_{x \to a} f(x)$ exists.

(9) If $\lim\limits_{x \to a} g(x)$ exists, $\lim\limits_{x \to a} f(g(x))$ exists.

#7 Find the value. ($[x]$; the greatest integer in x)

(1) Find the value of the limit: $\lim\limits_{x \to 2^+} \left(\dfrac{[x]^2 - 3[x] + 2}{[x] - 1}\right) \cdot \lim\limits_{x \to 2^-} \left(\dfrac{|x - 2|}{x - 2}\right)$

(2) Find the value of the limit: $\lim\limits_{x \to \infty} \dfrac{3}{x}\left[\dfrac{x}{2}\right]$

(3) Find the value of the limit: $\lim\limits_{x \to -\infty} (2x + \sqrt{[4x^2 + x]}\,)$

(4) For a function $f(x) = [x]^2 - a[x]$,

 find the value of the real number a so that $\lim\limits_{x \to 2} f(x)$ exists.

(5) Find the value of a positive integer a so that $\lim\limits_{x \to a} \frac{[x]^2+2x}{[x]}$ exists.

(6) For constants $a, b,$ and c, find the value of $a+b+c$ such that $\lim\limits_{x \to 1} \frac{x^3+ax+b}{(x-1)^2} = c$.

(7) For a cubic function $f(x)$ such that $\lim\limits_{x \to 0} \frac{f(x)}{x} = \lim\limits_{x \to 1} \frac{f(x)}{x-1} = 1$, find the value of $f(2)$.

(8) For two functions f and g such that $\lim\limits_{x \to \infty} f(x) = \infty$ and $\lim\limits_{x \to \infty} \{f(x) - 2g(x)\} = a$,

find the value of $\lim\limits_{x \to \infty} \frac{f(x)+2g(x)+3}{2f(x)-3g(x)-4}$. ($a$; constant)

(9) For a function $f(x) = x^3 + x^2 + x$, let $f^{-1}(x)$ be the inverse of $f(x)$.

Find the value of $\lim\limits_{x \to 0} \frac{f^{-1}(2x)}{x}$.

(10) For polynomial functions $f(x)$ and $g(x)$ such that $\lim\limits_{x \to 1} \frac{f(x)-1}{x-1} = 1$ and $\lim\limits_{x \to \infty} \frac{f(x)-x^3}{x^2+1} = 1$,

find the value of $f(2)$.

(11) For polynomial function $f(x)$ such that $\lim\limits_{x \to 0^+} \frac{x^3 f\left(\frac{1}{x}\right)-1}{x^3+x} = 5$ and $\lim\limits_{x \to 1} \frac{f(x)}{x^2+x-2} = \frac{1}{3}$,

find the value of $f(1)$.

(12) For polynomial function $g(x)$, $\lim\limits_{x \to 1} \frac{g(x)-2x}{x-1}$ exists. For a polynomial function $f(x)$ such

that $f(x) + x - 1 = (x-1)g(x)$, find the value of $\lim\limits_{x \to 1} \frac{f(x)g(x)}{x^2-1}$.

(13) When $\lim\limits_{x \to \infty} f(x) = \infty$ and $\lim\limits_{x \to a} \frac{\sqrt{g(x)}}{f(x)} = 2$, find the value of $\lim\limits_{x \to a} \frac{\log g(x)}{\log f(x)}$.

(14) For two polynomials f and g such that $\lim\limits_{x \to a} \frac{f(x)}{x-a} = 3$ and $\lim\limits_{x \to a} \frac{g(x)}{x-a} = 2$,

find the value of $\lim\limits_{x \to a} \frac{2f(x)+3g(x)}{f(x)-g(x)}$.

(15) For two real numbers x and y, a function $f(x)$ satisfies:

i) $f(x+y) = f(x) + f(y) + a$ and

ii) $\lim\limits_{x \to 2} \frac{f(x-2)}{x-2} = 1$

When $\lim\limits_{x \to 0} f(x) = f(0)$, find the value of the constant a.

(16) Find the value of a constant a at which $\lim\limits_{x \to 1} \frac{2x^2+a^2x-3a}{3x^2+a^2x-4a}$ does not exist.

#8 For two functions $f(x)$ and $g(x)$ such that

i) $x + f(x) = g(x)\{x - f(x)\}$ and

ii) $\lim_{x \to 0} g(x) = 2$,

determine if the limit exists.

(1) $\lim_{x \to 0} \dfrac{f(x)}{x}$

(2) $\lim_{x \to 0} f(x)$

(3) $\lim_{x \to 0} \dfrac{x^2 + f(x)}{x^2 - f(x)}$

#9 Find the limit.

(1) $\lim_{x \to 0} \dfrac{\sin 3x}{\sin 2x}$

(4) $\lim_{x \to 0} \dfrac{\tan 4x}{\tan 3x}$

(7) $\lim_{x \to 2\pi} \dfrac{\sin x}{x^2 - 4\pi^2}$

(2) $\lim_{x \to 0} \dfrac{\sin x^{\circ}}{x}$

(5) $\lim_{x \to 0} \dfrac{1 - \cos 4x}{x^2}$

(8) $\lim_{x \to 0} \dfrac{\cos 6x - 1}{2x}$

(3) $\lim_{x \to 0} \dfrac{\sin x - 2 \sin 2x}{x \cos x}$

(6) $\lim_{x \to 3} \dfrac{x - 3}{\sin \pi x}$

(9) $\lim_{x \to \infty} \dfrac{3^x}{3^x - 2^x}$

(10) $\lim_{x \to \infty} (3^x + 2^x)^{\frac{1}{x}}$

(15) $\lim_{x \to 0} \dfrac{\log(1+x)}{\tan x}$

(11) $\lim_{x \to \infty} \{\log(2 + 3x) - \log x\}$

(16) $\lim_{x \to 0} \dfrac{e^{2x} - 1}{\sin x}$

(12) $\lim_{x \to 3} \{\log|x^2 - 9| - \log|x - 3|\}$

(17) $\lim_{x \to 1} \dfrac{\sin\left(\cos\frac{\pi}{2}x\right)}{x - 1}$

(13) $\lim_{x \to 0} \dfrac{\log(1+x)}{x}$

(14) $\lim_{x \to 0} \dfrac{a^x - 1}{x} \quad (a > 0)$

(18) $\lim_{x \to \frac{\pi}{2}} \left(\dfrac{1}{\cos x} - \tan x\right)$

(19) $\lim_{x \to \infty} \left(1 + \sin\dfrac{1}{x}\right)^x$

(20) $\lim_{x \to 2} \{\log_4|x^2 - 4| - \log_4|x - \sqrt{x^2 + x - 2}|\}$

(21) $\lim_{x \to 0} \left\{\dfrac{\sin x \tan x}{x^2} + \dfrac{\sin 2x \tan 2x}{x^2} + \cdots\cdots + \dfrac{\sin 10x \tan 10x}{x^2}\right\}$

(22) $\lim_{x \to \infty} \sin\left(\tan\dfrac{1}{x}\right)\cot\dfrac{1}{x}$

(23) $\lim_{n \to \infty} \left\{\dfrac{1}{2}\left(1 + \dfrac{1}{n}\right)\left(1 + \dfrac{1}{n+1}\right)\cdots\cdots\left(1 + \dfrac{1}{n+n}\right)\right\}^n$

(24) $\lim_{x \to \infty} \left(\dfrac{x}{x-1}\right)^x \quad (x > 1)$

(28) $\lim_{x \to 0} \left(\dfrac{\sin x + \cos x}{\cos x}\right)^{\cos x}$

(25) $\lim_{x \to 0} \dfrac{1 - \cos 2x}{2 \sin^2 x}$

(29) $\lim_{x \to 0} \dfrac{x}{\sqrt[3]{e^x} - 1}$

(26) $\lim_{x \to -\infty} \dfrac{e^x + x^3 - 1}{1 + 2x^3}$

(30) $\lim_{x \to 2} \left(\dfrac{x}{2}\right)^{\frac{1}{x-2}}$

(27) $\lim_{x \to 0} \dfrac{\sin \pi(1-x)}{\ln(1+x)}$

(31) $\lim_{x \to 1} x^{\frac{1}{1-x}}$

#10 Find the values of constants a and b for the following limit.

(1) $\lim\limits_{x\to 0} \dfrac{ax\sin x+b}{1-\cos x} = 1$

(2) $\lim\limits_{x\to 0} \dfrac{\sin 2x}{\sqrt{ax+b}-1} = 2$

(3) $\lim\limits_{x\to 0} \dfrac{\tan x}{\sin(ax+b)} = \dfrac{1}{2}$ $\left(0 \le b < \dfrac{\pi}{2}\right)$

(4) $\lim\limits_{x\to\infty} \dfrac{2^{x+1}-3^{x+1}}{2^x-3^x} = a$; $\lim\limits_{x\to\infty} (4^x + 5^x)^{\frac{1}{x}} = b$

(5) $\lim\limits_{x\to 0} \dfrac{e^x \ln(x+a)}{b\sin x} = \dfrac{1}{2}$

(6) $\lim\limits_{x\to 1} \dfrac{\sin(x-1)}{\sqrt{ax-1}-b} = 1$

(7) $\lim\limits_{x\to\frac{\pi}{2}} \dfrac{ax^2-\pi^2}{\cos x} = b$ $(b \ne 0)$

#11 Find the value.

(1) For a continuous function $f(x)$ such that $(e^{4x} - 1)f(x) = \sin 2\pi x$, find the value of $f(0)$.

(2) $f(x)$ is a cubic function with $f(1) = 3$ and $f(2) = 12$. When $f(-x) = -f(x)$ for any real number x, find the value of $\lim\limits_{x\to 0} \dfrac{f(\sin x)}{\sin f(x)}$.

(3) For a sequence $\{a_n\}$ with $\lim\limits_{n\to\infty} \left(1 + \dfrac{3}{n}\right)^{a_n} = \dfrac{1}{e}$, find the value of $\lim\limits_{n\to\infty} \dfrac{a_n}{n}$.

(4) For a function $f(x)$ such that $\lim\limits_{x\to 0}\{f(x)\ln(1 + 3x)\} = 4$, find the value of $\lim\limits_{x\to 0} xf(x)$.

(5) For a function $f(x)$ such that $\lim\limits_{x\to 0} \dfrac{e^{2x}-1}{f(x)} = 5$, find the value of $\lim\limits_{x\to 0} \dfrac{f(x)}{x}$.

(6) For a function $f(x)$ such that $\lim\limits_{x\to\infty} x^2 f(x) = 2$, find the value of $\lim\limits_{x\to\infty} x^2 \ln\{1 + 2f(x)\}$.

(7) For a positive constant a such that $\lim\limits_{x\to 0} \dfrac{(a+2)^x-a^x}{x} = \ln 3$, find the value of a.

(8) For a function $f(x) = 2\ln(x + 1) + 1$, let the inverse function of $f(x)$ be $g(x)$. Find the value of $\lim\limits_{x\to 1} \dfrac{f(x-1)-f(0)}{g(x)-g(1)}$.

#12 For a function $f(x)$ such that $\lim\limits_{x\to 0} \dfrac{f(x)}{x} = 2$,

determine whether the following statements are true or false.

(1) $\lim\limits_{x\to 0} \dfrac{\tan x}{f(x)} = 1$

(2) $\lim\limits_{x\to 0} \dfrac{\tan f(x)}{f(x)} = 1$

(3) For a function $g(x)$, if $\lim\limits_{x\to 0} \dfrac{f(x)}{g(x)} = 1$, then $\lim\limits_{x\to 0} \dfrac{\tan f(x)}{\tan g(x)} = 1$.

#13 For a function $f(x) = \dfrac{b^x + \log_a x}{a^x + \log_b x}$ $(a > 0,\ b > 0,\ a \neq 1,\ b \neq 1)$,

determine whether the following statements are true or false.

(1) If $1 < a < b$, then $f(x) > 1$ for all x. $(x > 1)$

(2) If $b < a < 1$, then $\displaystyle\lim_{x\to\infty} f(x) = 0$.

(3) $\displaystyle\lim_{x\to 0^+} f(x) = \log_a b$

#14 For two functions $f(x)$ and $g(x)$ such that $\displaystyle\lim_{x\to\infty} f(x) = \infty$ and $\displaystyle\lim_{x\to\infty} g(x) = \infty$,

determine whether the following statements are true or false.

(1) If $\displaystyle\lim_{x\to\infty} \frac{g(x)}{f(x)} = 0$, then $\displaystyle\lim_{x\to\infty} \frac{e^{g(x)}}{e^{f(x)}} = 1$.

(2) If $\displaystyle\lim_{x\to\infty} \frac{g(x)}{f(x)} = 1$, then $\displaystyle\lim_{x\to\infty} \frac{\ln g(x)}{\ln f(x)} = 0$.

(3) If $\displaystyle\lim_{x\to\infty} \frac{g(x)}{f(x)} = 1$, then $\displaystyle\lim_{x\to\infty} \frac{\ln\left\{1 + \frac{1}{g(x)}\right\}}{\ln\left\{1 + \frac{1}{f(x)}\right\}} = 1$.

#15 For a function $f(x)$, determine whether the following statements are true or false.

(1) If $f(x) = x^2$, then $\displaystyle\lim_{x\to 0} \frac{e^{f(x)} - 1}{x} = 0$.

(2) If $\displaystyle\lim_{x\to 0} \frac{e^x - 1}{f(x)} = 1$, then $\displaystyle\lim_{x\to 0} \frac{2^x - 1}{f(x)} = \ln 2$

(3) If $\displaystyle\lim_{x\to 0} f(x) = 0$, then $\displaystyle\lim_{x\to 0} \frac{e^{f(x)} - 1}{x}$ exists.

#16 For a function $f(x)$ such that $\displaystyle\lim_{x\to 0} \frac{f(x)}{\ln(1+x)} = 1$, find the limit.

(1) $\displaystyle\lim_{x\to 0} \frac{\sin x}{f(x)}$

(2) $\displaystyle\lim_{x\to 0} \frac{f(x) + x}{\ln(1+x)}$

(3) $\displaystyle\lim_{x\to 0} \frac{\{f(x)\}^2}{\ln(1+x)}$

#17 At what points is the function discontinuous?

(1) $f(x) = \dfrac{x-1}{|x-1|} x^2$

(2) $f(x) = x^2 + \dfrac{x^2}{1+x^2} + \dfrac{x^2}{(1+x^2)^2} + \cdots\cdots$

#18 Given the piecewise-defined function $f(x)$ defined below,

identify any value(s) of x at which $f(x)$ is discontinuous and describe the discontinuity exhibited.

(1) $f(x) = \begin{cases} x^2 - 2x + 4, & x \le 2 \\ x^3 - 5, & x > 2 \end{cases}$

(2) $f(x) = \begin{cases} x, & x \text{ is a rational number in } [0, 1] \\ 1 - x, & x \text{ is an irrational number in } [0, 1] \end{cases}$

(3) $f(x) = \begin{cases} \dfrac{x}{\sqrt{1+x}-1}, & x \ne 0 \\ 1, & x = 0 \end{cases}$

(4) $f(x) = \begin{cases} \dfrac{[x]}{x}, & x \ne 0 \\ 1, & x = 0 \end{cases}$ ($[x]$ is the greatest integer in x)

#19 Examine for continuity of the function.

(1) $f(x) = \begin{cases} x \sin \dfrac{1}{x}, & x \ne 0 \\ 0, & x = 0 \end{cases}$

(2) $f(x) = \lim\limits_{n \to \infty} \dfrac{1}{1+(-x^2+5)^{2n}}$

(3) $f(x) = \lim\limits_{n \to \infty} \dfrac{1-\{\log_2(1+|x|)\}^n}{1+\{\log_2(1+|x|)\}^n}$

(4) $f(x) = \lim\limits_{n \to \infty} \dfrac{x^n+2x+1}{x^{n-1}+1}$

(n; positive integer)

#20 Show the indicated function has at least one real number solution in the interval.

(1) $x^4 + x^3 - 5x + 1 = 0$, $(1, 3)$

(2) $\sin x = x \cos x$, $\left(\pi, \dfrac{3}{2}\pi \right)$

#21 Find the values of constants a and b:

(1) When a function $f(x) = \begin{cases} \dfrac{a\sqrt{x+1}-b}{x-1}, & x > 1 \\ 2x - 1, & x \le 1 \end{cases}$ is continuous at $x = 1$.

(2) When a function $f(x) = \lim\limits_{n \to \infty} \dfrac{x^{2n+1}+ax+b}{x^{2n}+1}$ is continuous at all real number x.

(3) When a function $f(x) = [x]^2 + (ax + b)[x]$ is continuous at all real number x.

    ($[x]$ is the greatest integer in x.)

(4) When a function $f(x) = \begin{cases} x(x - 1), & |x| > 1 \\ -x^2 + ax + b, & |x| \le 1 \end{cases}$ is continuous at all real number x.

(5) For a function $f(x) = x^2 - 4x + a$ and a function $g(x) = \lim\limits_{n \to \infty} \dfrac{2|x-b|^n+1}{|x-b|^n+1}$,

    $h(x) = f(x)g(x)$.

    When the function $h(x)$ is continuous at all real number x, find the values of a and b.

#22 Find the value of the constant a.

(1) For real number, $f(x) = \begin{cases} \frac{\sin 2(x-1)}{x-1}, & |x| \neq 1 \\ a, & |x| = 1 \end{cases}$.

When $f(x)$ is continuous at $x = 1$, find the value of a.

(2) When $f(x) = \begin{cases} \frac{e^{3x}-1}{\sin 2x}, & -\frac{\pi}{2} \leq x < 0, \ 0 < x \leq \frac{\pi}{2} \\ a, & x = 0 \end{cases}$ is continuous at $x = 0$,

find the value of a.

(3) For two functions $f(x) = \begin{cases} x^2 + 2x + 2, & x \geq 1 \\ x + 1, & x < 1 \end{cases}$ and $g(x) = |x - 2a|$,

a function $h(x) = (g \circ f)(x)$ is continuous at all real number x. Find the value of a.

(4) When a function defined by $f(x) = \begin{cases} x^2 + \frac{x^2}{1+\tan^2 x} + \frac{x^2}{(1+\tan^2 x)^2} + \cdots\cdots, & x \neq 0 \\ a, & x = 0 \end{cases}$ is continuous

at $x = 0$ in the interval $\left(-\frac{\pi}{2}, \frac{\pi}{2}\right)$, find the value of the constant a.

#23 Find the value.

(1) For any real number x, the continuous function $f(x)$ such that $f(x + 4) = f(x)$ is defined

by $f(x) = \begin{cases} 3x, & 0 \leq x < 1 \\ x^2 + ax + b, & 1 \leq x \leq 4 \end{cases}$. Find the value of $f(10)$.

(2) When a quadratic function $f(x)$ has the leading coefficient 1, and two other functions

$g(x)$ and $h(x)$ are defined by $g(x) = \lim\limits_{n \to \infty} \frac{x^{2n-1}-1}{x^{2n}+1}$, $h(x) = \begin{cases} \frac{|x|}{x}, & x \neq 0 \\ 0, & x = 0 \end{cases}$,

$f(x)g(x)$ and $f(x)h(x)$ are continuous. Find the value of $f(10)$.

(3) For two functions $f(x) = -x^2 + 4x - 2$ and $(x) = 2\sin\frac{x}{2}$, let the maximum and minimum

values of $(f \circ g)(x)$ in the interval $[0, 2\pi]$ be M and n, respectively.

Find the value of $M + m$.

(4) For $x > 0$, the function $f(x)$ such that $(4x - 1)f(x) = \ln 4x$ is continuous.

Find the value of $f\left(\frac{1}{4}\right)$.

#24 For a continuous function $f(x)$ in $[a, b]$ with $f(a) = b$ and $f(b) = a$,

show that there is c $(a < c < b)$ such that $f(c) = c$.

#25 For a function $f(x) = |x| + \dfrac{|x|}{1+|x|} + \dfrac{|x|}{(1+|x|)^2} + \dfrac{|x|}{(1+|x|)^3} + \cdots\cdots$,

determine whether the following statements are true or false.

(1) $\displaystyle\lim_{x\to 0} f(\sin x) = f(\lim_{x\to 0} \sin x)$

(2) $\displaystyle\lim_{h\to 0} \dfrac{f(h)-f(-h)}{h}$ exists.

(3) For any real number a, $\displaystyle\lim_{h\to 0} f(a+h) = \lim_{h\to 0} f(a-h)$

#26 For two functions $f(x)$ and $g(x)$, determine whether the following statements are true or false.

(1) If $\displaystyle\lim_{x\to 0} f(x)$ and $\displaystyle\lim_{x\to 0} g(x)$ do not exist, $\displaystyle\lim_{x\to 0} \{f(x) + g(x)\}$ does not exist.

(2) If $y = f(x)$ is continuous at $x = 0$, then $y = |f(x)|$ is also continuous at $x = 0$.

(3) If $y = |f(x)|$ is continuous at $x = 0$, then $y = f(x)$ is also continuous at $x = 0$.

(4) When $f(x) = \begin{cases} 1 & , \ x \geq 0 \\ -1 & , \ x < 0 \end{cases}$ and $g(x) = |x|$, $(g \circ f)(x)$ is continuous at $x = 0$.

(5) If $(g \circ f)(x)$ is continuous at $x = 0$, then $f(x)$ is continuous at $x = 0$.

(6) If $(f \circ f)(x)$ is continuous at $x = 0$, then $f(x)$ is continuous at $x = 0$.

Chapter 3. The Derivative

3-1 Definition of the Derivative

1. Notation of a Tangent

Let f be any function, and let $P = (x_0, f(x_0))$ be any point of the graph of f.

Consider the equation of the line tangent to the graph at P.

Note that the line passes through the given point P, and any non-vertical line through P has an equation of the form $y - f(x_0) = m(x - x_0)$, where m is the slope of the line.

When we take the line through the given point $P = (x_0, f(x_0))$ on a curve and a nearby movable point $Q = (x, f(x))$ on that curve, the line through P and Q is called a *secant line*.

The slope of the secant line is

$$m_{sec} = \frac{\text{Difference of } y-\text{values}}{\text{Difference of } x-\text{values}} = \frac{f(x)-f(x_0)}{x-x_0}, \ x \neq x_0$$

As the point Q approaches P, the secant line approaches the tangent line, and therefore, the slope of the secant line should approach the slope of the tangent line.

That is, $m_{tan} = \lim\limits_{x \to x_0} m_{sec}$

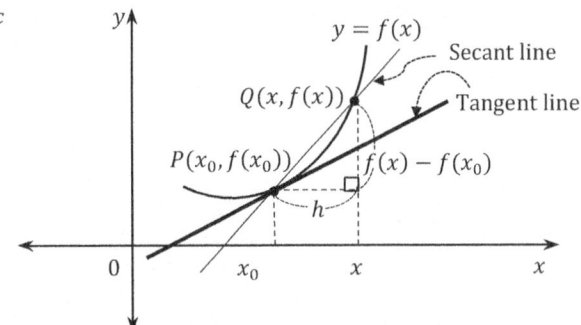

The ratio $\boxed{\dfrac{\Delta f}{\Delta x} = \dfrac{f(x_0+h)-f(x_0)}{h} = \dfrac{f(x)-f(x_0)}{x-x_0}, \ \Delta x = x - x_0 = h \neq 0,}$ is called the *difference*

quotient of f at x_0.

The *tangent line* to the curve $y = f(x)$ at the point $P(x_0, f(x_0))$, if not vertical, is the line

through P with slope m satisfying $\boxed{m = \lim\limits_{h \to 0} \dfrac{f(x_0+h)-f(x_0)}{h} = \lim\limits_{x \to x_0} \dfrac{f(x)-f(x_0)}{x-x_0}}$

Example   Find the equation of the line tangent to the curve $y = x^2$ at the point (x_0, y_0).

Note that: $f(x_0) = x_0{}^2$ and $f(x_0 + h) = (x_0 + h)^2 = x_0{}^2 + 2x_0 h + h^2$ $(h \neq 0)$

$$m = \lim_{h \to 0} \frac{f(x_0+h)-f(x_0)}{h} = \lim_{h \to 0} \frac{x_0{}^2+2x_0h+h^2-x_0{}^2}{h} = \lim_{h \to 0} \frac{2x_0h+h^2}{h} = \lim_{h \to 0}(2x_0 + h) = 2x_0 + 0 = 2x_0$$

∴ The equation of the tangent line is $y - y_0 = 2x_0(x - x_0)$

Example Find the equation of the line tangent to the curve $y = x^2 - x + 1$ at the point $(0, 1)$.

Let $x_0 = 0$.

Then, $f(x_0) = f(0) = 1$

$f(x_0 + h) = f(h) = h^2 - h + 1 \ (h \neq 0)$

$m = \lim\limits_{h \to 0} \dfrac{f(x_0 + h) - f(x_0)}{h} = \lim\limits_{h \to 0} \dfrac{h^2 - h + 1 - 1}{h} = \lim\limits_{h \to 0} (h - 1) = 0 - 1 = -1$

∴ The equation of the tangent line at the point $(0, 1)$ is $y - 1 = -1(x - 0)$; i.e., $y = -x + 1$

2. Average Velocity and Instantaneous Velocity

If we drive a car from one town to another 100 miles away in 2 hours, our average velocity is 50 miles per hour. *Average velocity* is the distance from the first position to the second position divided by the elapsed time.

Let an object P moves along a coordinate line so that its position at time t is given by $y = f(t)$. At time t_0, the object is at $f(t_0)$; at the nearby time $t_0 + h$, it is at $f(t_0 + h)$.

Thus, the object has moved $f(t_0 + h) - f(t_0)$ units of distance in $h \ (h > 0)$ units of time.

Therefore, the average velocity during the time interval h is the ratio $v_{ave} = \dfrac{f(t_0 + h) - f(t_0)}{h}$.

However, during our trip the speedometer reading was often different from the average velocity. Since the average velocity describes what happens in an interval of time, it is an interval property.

Now, consider a small interval of time $[t_0, \ t_0 + h]$, where $h > 0$, and the average velocity v_{ave} during this interval.

For given distance y as a function f of t, the *instantaneous velocity* at t_0 is defined to be

$$v = \lim\limits_{h \to 0} v_{ave} = \lim\limits_{h \to 0} \dfrac{f(t_0 + h) - f(t_0)}{h} \quad \text{provided this limit exists.}$$

Note that instantaneous velocity is a point property; it is a limit.

Example An object P moves vertically in a straight line under the following law of motion:

$$y = 8t - t^2, \text{ where } t \text{ is in seconds and } y \text{ is in feet.}$$

Find (1) the velocity at any time t.

(2) the domain of values of $t > 0$ for which velocity is positive.

(3) maximum value of y.

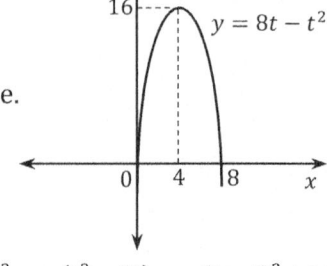

$$y = 8t - t^2 = -(t^2 - 8t) = -(t - 4)^2 + 16$$

(1) $v = \lim\limits_{h \to 0} \dfrac{f(t_0+h)-f(t_0)}{h} = \lim\limits_{h \to 0} \dfrac{\{8(t_0+h)-(t_0+h)^2\}-(8t_0-t_0{}^2)}{h}$

$= \lim\limits_{h \to 0} \dfrac{8t_0+8h-\left(t_0{}^2+2t_0h+h^2\right)-8t_0+t_0{}^2}{h} = \lim\limits_{h \to 0} \dfrac{8h-2t_0h-h^2}{h} = \lim\limits_{h \to 0}(8 - 2t_0 - h)\ (h \neq 0)$

$= 8 - 2t_0$ Feet/second

(2) Since $v = 8 - 2t > 0,\ \ t < 4$

Since $t > 0,\ \ 0 < t < 4$

(3) Note that the velocity v is positive (going up) when $0 < t < 4$ and negative (coming down) when $4 < t < 8$.

When $t = 4$, we have the maximum height of y.

$\therefore$ Maximum value of y is $8 \cdot 4 - 4^2 = 32 - 16 = 16$ feet

3. Derivative

Note that the slope m of the tangent to the curve $y = f(x)$ at a point x is $m = \lim\limits_{h \to 0} \dfrac{f(x+h)-f(x)}{h}$.

When $y = f(t)$ relates distance and time, the velocity at t is $v = \lim\limits_{h \to 0} \dfrac{f(t+h)-f(t)}{h}$.

That is, the slope of tangent line and instantaneous velocity are obvious manifestations of the same basic idea.

Now add it to function and limit, we define the derivative.

Let f be defined on a neighborhood of x.

The *derivative* of the function f at the point x is the real number

$\lim\limits_{h \to 0} \dfrac{f(x+h)-f(x)}{h} = \lim\limits_{x \to a} \dfrac{f(x)-f(a)}{x-a}$

provided this limit exists.

We say the function is *differentiable* at x.

Finding a derivative is called *differentiations*.

We define another function f' given by the rule $f' : x \longmapsto \lim\limits_{h \to 0} \dfrac{f(x+h)-f(x)}{h}$.

f' is called the *derivative* of f.

That is, the value of $f'(x) = \lim\limits_{h \to 0} \dfrac{f(x+h)-f(x)}{h}$ is the derivative of f at x.

For example, if $f(x) = x^2$, then $f'(x_0) = 2x_0$

Note: $\left| f'(x) = \lim\limits_{h \to 0} \dfrac{f(x+h)-f(x)}{h} = \lim\limits_{a \to 0} \dfrac{f(x+a)-f(x)}{a} = \lim\limits_{b \to 0} \dfrac{f(x+b)-f(x)}{b} = \lim\limits_{x \to a} \dfrac{f(x)-f(a)}{x-a} \right.$

Note:

A point x is in the domain of f' $\underset{\text{if and only if}}{\Longleftrightarrow}$ $\begin{cases} f \text{ is defined on a neighborhood of } x \text{, and} \\ \lim\limits_{h\to 0} \dfrac{f(x+h)-f(x)}{h} \text{ exists.} \end{cases}$

If f is differentiable at every point in its domain, then f is called a differentiable function.

Thus, if f is a differentiable function, then f and f' have the same domain.

Caution!

The limit does not always exist.

That is, there are functions that do not have a derivative at some points where they are defined.

For example, consider the function $f(x) = |x|$.

$$\frac{f(x)-f(0)}{x-0} = \frac{|x|}{x} = \begin{cases} 1, & x > 0 \\ -1, & x < 0 \end{cases}$$

$\therefore$ $\lim\limits_{x\to 0} \dfrac{f(x)-f(0)}{x-0}$ does not tend to any single number.

$\therefore$ $f'(0)$ does not exist.; i.e., f is not differentiable at $x = 0$; f' is not defined at $x = 0$.

Example

(1) For a function $f(x) = \dfrac{1}{x}$, $x \neq 0$, compute the derivative $f'(x_0)$.

(2) Compute the derivative of the general quadratic function $f(x) = ax^2 + bx + c$, where a, b, and c are constants.

(1) $\dfrac{f(x)-f(x_0)}{x-x_0} = \dfrac{\frac{1}{x}-\frac{1}{x_0}}{x-x_0} = \dfrac{\frac{x_0-x}{xx_0}}{x-x_0} = \dfrac{x_0-x}{xx_0(x-x_0)} = \dfrac{-1}{xx_0}$

$\therefore$ $f'(x_0) = \lim\limits_{x\to x_0} \dfrac{f(x)-f(x_0)}{x-x_0} = \lim\limits_{x\to x_0} \dfrac{-1}{xx_0} = \dfrac{-1}{x_0{}^2}$

(2) $\dfrac{f(x)-f(x_0)}{x-x_0} = \dfrac{ax^2+bx+c-(ax_0{}^2+bx_0+c)}{x-x_0} = \dfrac{a(x^2-x_0{}^2)+b(x-x_0)}{x-x_0} = a(x+x_0) + b, \quad x \neq x_0$

$\therefore$ $f'(x_0) = \lim\limits_{x\to x_0} \dfrac{f(x)-f(x_0)}{x-x_0} = \lim\limits_{x\to x_0} \{a(x+x_0) + b\} = a(x_0+x_0) + b = 2ax_0 + b$

3-2 Continuity of Differentiable Functions

Differentiability implies continuity.

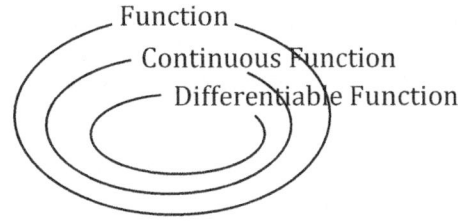

Differentiable function 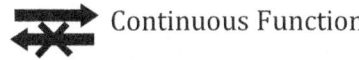 Continuous Function

---

> **If f' exists, then f is continuous at x.**

∵ Note that $f(x) = f(a) + \frac{f(x)-f(a)}{x-a}(x-a),\ x \neq a$

∴ $\lim_{x \to a} f(x) = \lim_{x \to a}\left\{f(a) + \frac{f(x)-f(a)}{x-a}(x-a)\right\} = \lim_{x \to a} f(a) + \lim_{x \to a}\frac{f(x)-f(a)}{x-a} \cdot \lim_{x \to a}(x-a)$

$\qquad\qquad = f(a) + f'(a) \cdot 0 = f(a)$

That is, $\lim_{x \to a} f(x) = f(a)$

Therefore, f is continuous at x.

However, the converse does not hold true.

For example, the absolute value function $f(x) = |x|$ is continuous but not differentiable at $x = 0$.

When $x = 0,\ \lim_{x \to 0} f(x) = \lim_{x \to 0}|x| = 0$ and $f(0) = 0$

∴ $\lim_{x \to 0} f(x) = f(0)$

Thus, $f(x)$ is continuous at $x = 0$.

But, $\lim_{h \to 0^-} \frac{f(0+h)-f(0)}{h} = \lim_{h \to 0^-} \frac{|h|}{h} = -1$

$\qquad \lim_{h \to 0^+} \frac{f(0+h)-f(0)}{h} = \lim_{h \to 0^+} \frac{|h|}{h} = 1$

That is, $\lim_{h \to 0^-} \frac{f(0+h)-f(0)}{h} \neq \lim_{h \to 0^+} \frac{f(0+h)-f(0)}{h}$

∴ $f'(x) = \lim_{h \to 0} \frac{f(0+h)-f(0)}{h}$ does not exist.

Therefore, the function $f(x) = |x|$ is not differentiable at $x = 0$.

In general, the derivative of a function does not exist at points of discontinuity and sharp points of the function.

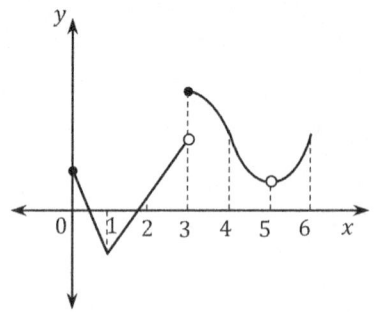

The graph of $y = f(x)$ in the open interval $(0, 6)$ is discontinuous at $x = 3$ and $x = 5$, and is not differentiable at $x = 1$, $x = 3$, and $x = 5$.

3-3 Computation of Derivatives

1. Rules for Finding Derivatives

From the definition of the derivative, we can find the derivative of a function directly.

That is, set the difference quotient $\dfrac{f(x+h)-f(x)}{h}$ and evaluate the limit.

(1) The constant and Power Rules

Let f and g be two functions which are differentiable at x.

Constant function $f(x) = c$ is a horizontal line which has slope zero everywhere.

1) Constant Function Rules

If $f(x) = c$, where c is a constant, then for any x, $f'(x) = 0$.

$\because\ f'(x) = \lim_{h\to 0} \frac{f(x+h)-f(x)}{h} = \lim_{h\to 0} \frac{c-c}{h} = \lim_{h\to 0} 0 = 0$

For example, $f(x) = 3\ \Rightarrow\ f'(x) = 0$; $f(x) = 2014\ \Rightarrow\ f'(x) = 0$

2) Identity Function Rules

The graph of $f(x) = x$ is a line through the origin with slope 1.

If $f(x) = x$, then $f'(x) = 1$.

$\because\ f'(x) = \lim_{h\to 0} \frac{f(x+h)-f(x)}{h} = \lim_{h\to 0} \frac{x+h-x}{h} = \lim_{h\to 0} \frac{h}{h} = \lim_{h\to 0} 1 = 1$

3) Power Rules

$f(x) = x^{-n}$
$\Rightarrow f'(x) = -nx^{-n-1}$

If $f(x) = x^n$, where n is a positive integer, then $f'(x) = nx^{n-1}$

Note that $\boxed{y = \sqrt{x}\ \Rightarrow\ y' = \dfrac{1}{2\sqrt{x}}\ ;\ y = \sqrt{f(x)}\ \Rightarrow\ y' = \dfrac{f'(x)}{2\sqrt{f(x)}}}$

$\because\ y = \sqrt{x} = x^{\frac{1}{2}}$

$\therefore\ y' = \frac{1}{2}x^{\frac{1}{2}-1} = \frac{1}{2}x^{-\frac{1}{2}} = \frac{1}{2}\frac{1}{\sqrt{x}} = \frac{1}{2\sqrt{x}}$

Note that $(a+b)^2 = a^2 + 2ab + b^2$
$$(a+b)^3 = a^3 + 3a^2b + 3ab^2 + b^3$$
$$(a+b)^4 = a^4 + 4a^3b + 6a^2b^2 + 4ab^3 + b^4$$
$$\vdots$$
$$(a+b)^n = a^n + na^{n-1}b + \frac{n(n-1)}{2}a^{n-2}b^2 + \cdots + nab^{n-1} + b^n$$

(2) Linear Operator

1) Constant Multiple Rule

> If c is a constant and f is a differentiable function, then $(cf)'(x) = c \cdot f'(x)$

$\because$ Let $F(x) = c \cdot f(x)$

Then, $F'(x) = \lim_{h \to 0} \frac{F(x+h)-F(x)}{h} = \lim_{h \to 0} \frac{c \cdot f(x+h)-c \cdot f(x)}{h} = \lim_{h \to 0} c \cdot \frac{f(x+h)-f(x)}{h}$

$$= c \cdot \lim_{h \to 0} \frac{f(x+h)-f(x)}{h} = c \cdot f'(x)$$

$\therefore (c \cdot f)'(x) = c \cdot f'(x)$

For example, $y = 3x \Rightarrow y' = 3(x)' = 3 \cdot (1x^{1-1}) = 3(x^0) = 3 \cdot 1 = 3$

$$y = -5x^4 \Rightarrow y' = -5(x^4)' = -5 \cdot (4x^{4-1}) = -5 \cdot (4x^3) = -20x^3$$

2) Sum Rule

> If f and g are differentiable functions, then the sum $f+g$ is differentiable function and
> $(f+g)'(x) = f'(x) + g'(x)$

$\because$ Let $F(x) = f(x) + g(x)$

Then, $F'(x) = \lim_{h \to 0} \frac{F(x+h)-F(x)}{h} = \lim_{h \to 0} \frac{f(x+h)+g(x+h)-\{f(x)+g(x)\}}{h}$

$$= \lim_{h \to 0}\left[\left\{\frac{f(x+h)-f(x)}{h}\right\} + \left\{\frac{g(x+h)-g(x)}{h}\right\}\right] = \lim_{h \to 0}\frac{f(x+h)-f(x)}{h} + \lim_{h \to 0}\frac{g(x+h)-g(x)}{h}$$

$$= f'(x) + g'(x)$$

3) Difference Rule

> If f and g are differentiable functions, then the difference $f-g$ is differentiable function
> and $(f-g)'(x) = f'(x) - g'(x)$

$\because \{f(x)-g(x)\}' = \{f(x)+(-1)g(x)\}'$

$\quad\quad = f'(x) + \{(-1)g(x)\}'$ By sum rule

$\quad\quad = f'(x) + (-1)g'(x)$ By constant multiple rule

$\quad\quad = f'(x) - g'(x)$

For example, $y = 2x^4 - 5x^3 + 4x^2 - 6$

$$\Rightarrow \ y' = (2x^4)' - (5x^3)' + (4x^2)' - (6)'$$

$$= 8x^3 - 15x^2 + 8x$$

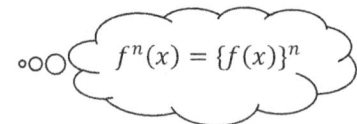

The derivative of a product of functions is not equal to the product of the derivatives of the functions.

$$(fg)' \neq f'g'$$

(3) Product and Quotient Rules

1) Product Rule

① If f and g are differentiable functions, then the product fg is differentiable and

$$(f \cdot g)'(x) = f'(x) \cdot g(x) + f(x) \cdot g'(x)$$

Note: If f, g, and h are differentiable functions,

$$(f \cdot g \cdot h)'(x) = f'(x) \cdot g(x) \cdot h(x) + f(x) \cdot g'(x) \cdot h(x) + f(x) \cdot g(x) \cdot h'(x)$$

② If f is differentiable and n is a positive integer, then

$$[\{f(x)\}^n]' = n\{f(x)\}^{n-1} \cdot f'(x)$$

$f^n(x) = \{f(x)\}^n$

For example,

$$y = (x-3)(2x-1) \ \Rightarrow \ y' = (x-3)'(2x-1) + (x-3)(2x-1)'$$

$$= 1 \cdot (2x-1) + (x-3) \cdot 2$$

$$= 2x - 1 + 2x - 6$$

$$= 4x - 7$$

$$y = \frac{x^3}{(x+1)^2} \ \Rightarrow \ y' = x^3(x+1)^{-2} = \{x^3\}'(x+1)^{-2} + x^3\{(x+1)^{-2}\}'$$

$$= (3x^2)(x+1)^{-2} + x^3\{-2(x+1)^{-3}\}(x+1)'$$

$$= \frac{3x^2}{(x+1)^2} + x^3 \cdot \left\{\frac{-2}{(x+1)^3}\right\} \cdot 1$$

$$= \frac{3x^2}{(x+1)^2} - \frac{2x^3}{(x+1)^3}$$

$$= \frac{3x^2(x+1) - 2x^3}{(x+1)^3} = \frac{x^2(x+3)}{(x+1)^3}$$

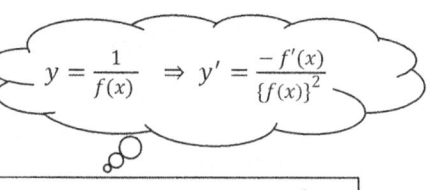

$$y = \frac{1}{f(x)} \ \Rightarrow \ y' = \frac{-f'(x)}{\{f(x)\}^2}$$

2) Quotient Rule

If f and g are differentiable functions with $g(x) \neq 0$, then the quotient function $\frac{f}{g}$ is differentiable and $\left(\frac{f}{g}\right)'(x) = \frac{f'(x)g(x) - f(x)g'(x)}{\{g(x)\}^2}$

For example, $y = \frac{x^3}{(x+1)^2} \ \Rightarrow \ y' = \frac{\{x^3\}'(x+1)^2 - x^3\{(x+1)^2\}'}{\{(x+1)^2\}^2} = \frac{3x^2(x+1)^2 - x^3\{2(x+1)\}}{(x+1)^4}$

$$= \frac{3x^2(x+1) - 2x^3}{(x+1)^3} = \frac{x^3 + 3x^2}{(x+1)^3} = \frac{x^2(x+3)}{(x+1)^3}$$

(4) The Chain Rule

The computation of the derivative of a composite function can be accomplished.

Let f and g be real functions.

> If f is differentiable at x, and g is differentiable at $f(x)$.
>
> Then, the composite function $g \circ f$ is differentiable at x and $(g \circ f)'(x) = g'(f(x)) \cdot f'(x)$

For example, $y = \left(\frac{x}{x^2+1}\right)^3 \Rightarrow y' = 3\left(\frac{x}{x^2+1}\right)^2 \left(\frac{x}{x^2+1}\right)' = 3\left(\frac{x}{x^2+1}\right)^2 \left\{\frac{1(x^2+1)-x(2x)}{(x^2+1)^2}\right\}$

$$= 3\left(\frac{x}{x^2+1}\right)^2 \left\{\frac{1-x^2}{(x^2+1)^2}\right\} = \frac{3x^2(1-x^2)}{(x^2+1)^4}$$

2. Derivatives of the Trigonometric Functions

Note that the functions $f(x) = \sin x$ and $g(x) = \cos x$ are both differentiable.

Note: $\sin(0) = 0$ and $\displaystyle\lim_{x \to 0} \frac{\sin x}{x} = 1$

The definition of derivative gives $\sin'(0) = \displaystyle\lim_{\theta \to 0} \frac{\sin\theta - \sin 0}{\theta - 0} = \lim_{\theta \to 0}\frac{\sin\theta}{\theta} = 1$

(1) The derivative Formulas

1) $\boxed{y = \sin x \Rightarrow y' = \cos x}$

$\because\ y' = \displaystyle\lim_{h\to 0}\frac{\sin(x+h)-\sin x}{h} = \lim_{h\to 0}\frac{2\cos\left(x+\frac{h}{2}\right)\sin\frac{h}{2}}{h} = \lim_{h\to 0}\cos\left(x+\frac{h}{2}\right)\cdot\frac{\sin\frac{h}{2}}{\frac{h}{2}} = \cos x$

2) $\boxed{y = \cos x \Rightarrow y' = -\sin x}$

$\because\ y' = \displaystyle\lim_{h\to 0}\frac{\cos(x+h)-\cos x}{h} = \lim_{h\to 0}\frac{-2\sin\left(x+\frac{h}{2}\right)\sin\frac{h}{2}}{h} = -\lim_{h\to 0}\sin\left(x+\frac{h}{2}\right)\cdot\frac{\sin\frac{h}{2}}{\frac{h}{2}} = -\sin x$

Note: $(\cos x)' = \left\{\sin\left(x+\frac{\pi}{2}\right)\right\}' = \cos\left(x+\frac{\pi}{2}\right) = -\sin x$

3) $\boxed{y = \tan x \Rightarrow y' = \sec^2 x}$

$\because\ y' = \left(\frac{\sin x}{\cos x}\right)' = \frac{(\sin x)'\cos x - \sin x(\cos x)'}{\cos^2 x} = \frac{(\cos x)\cos x - \sin x(-\sin x)}{\cos^2 x} = \frac{\cos^2 x + \sin^2 x}{\cos^2 x}$

$= \frac{1}{\cos^2 x} = \sec^2 x$

4) $\boxed{y = \cot x \Rightarrow y' = -\csc^2 x}$

$\because\ y' = \left(\frac{\cos x}{\sin x}\right)' = \frac{(\cos x)'\sin x - \cos x(\sin x)'}{\sin^2 x} = \frac{(-\sin x)\sin x - \cos x(\cos x)}{\sin^2 x} = \frac{-\sin^2 x - \cos^2 x}{\sin^2 x}$

$= \frac{-1}{\sin^2 x} = -\csc^2 x$

5) $\boxed{y = \sec x \;\Rightarrow\; y' = \sec x \cdot \tan x}$

$$\because\; y' = \left(\frac{1}{\cos x}\right)' = \frac{(1)' \cos x - 1(\cos x)'}{\cos^2 x} = \frac{0 \cdot \cos x - (-\sin x)}{\cos^2 x} = \frac{\sin x}{\cos^2 x}$$

$$= \frac{1}{\cos x} \cdot \frac{\sin x}{\cos x} = \sec x \cdot \tan x$$

6) $\boxed{y = \csc x \;\Rightarrow\; y' = -\csc x \cdot \cot x}$

$$\because\; y' = \left(\frac{1}{\sin x}\right)' = \frac{(1)' \sin x - 1(\sin x)'}{\sin^2 x} = \frac{0 \cdot \sin x - (\cos x)}{\sin^2 x} = \frac{-\cos x}{\sin^2 x}$$

$$= \frac{-1}{\sin x} \cdot \frac{\cos x}{\sin x} = -\csc x \cdot \cot x$$

Note: <u>Derivative of Trigonometric Functions</u>

$$\boxed{\begin{aligned}
y &= \sin x \;\Rightarrow\; y' = \cos x \\
y &= \cos x \;\Rightarrow\; y' = -\sin x \\
y &= \tan x \;\Rightarrow\; y' = \sec^2 x \\
y &= \cot x \;\Rightarrow\; y' = -\csc^2 x \\
y &= \sec x \;\Rightarrow\; y' = \sec x \cdot \tan x \\
y &= \csc x \;\Rightarrow\; y' = -\csc x \cdot \cot x
\end{aligned}}$$

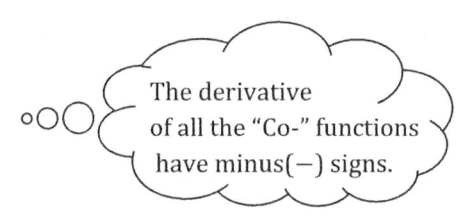

The derivative of all the "Co-" functions have minus$(-)$ signs.

Example Find the derivative of each of the following functions.

$$(1)\; y = (\sin x + \cos x)^3 \qquad (2)\; y = \frac{1+\sin x}{\cos x}$$

$(1)\; y' = \{(\sin x + \cos x)^3\}' = 3(\sin x + \cos x)^2 (\sin x + \cos x)'$

$\qquad = 3(\sin x + \cos x)^2 (\cos x - \sin x)$

$(2)\; y' = \left(\dfrac{1+\sin x}{\cos x}\right)' = \dfrac{(1+\sin x)' \cos x - (1+\sin x)(\cos x)'}{\cos^2 x} = \dfrac{\cos x \cos x - (1+\sin x)(-\sin x)}{\cos^2 x}$

$\qquad = \dfrac{\cos^2 x + \sin x + \sin^2 x}{\cos^2 x} = \dfrac{1+\sin x}{\cos^2 x} = \dfrac{1+\sin x}{1-\sin^2 x} = \dfrac{1+\sin x}{(1+\sin x)(1-\sin x)} = \dfrac{1}{1-\sin x}$

(2) Leibniz Notation

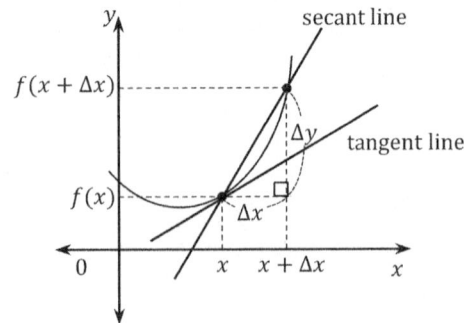

The ratio $\dfrac{\Delta y}{\Delta x} = \dfrac{f(x+\Delta x) - f(x)}{\Delta x}$ is the slope of a secant line through $(x, f(x))$.

As $\Delta x \to 0$, the slope of this secant line approaches that of the tangent line.

Leibniz denoted derivatives by the symbol $\frac{df}{dx}$, rather than $f'(x)$.

$\therefore \frac{df}{dx} = \lim_{\Delta x \to 0} \frac{\Delta y}{\Delta x} = \lim_{\Delta x \to 0} \frac{f(x+\Delta x)-f(x)}{\Delta x} = \lim_{x \to x_0} \frac{f(x)-f(x_0)}{x-x_0}$

For example, if $f(x) = 3x^2 + 2x$, then $\frac{df}{dx} = \frac{d}{dx}(3x^2 + 2x) = \frac{d}{dx}(3x^2) + \frac{d}{dx}(2x) = 6x + 2$

There are many variations on this, chosen for convenience.

The Chain Rule

Suppose $y = f(u)$ and $u = g(x)$.

In Leibniz notation, the chain rule takes the form:

$\frac{dy}{dx} = \frac{dy}{du} \cdot \frac{du}{dx}$

$\frac{dy}{dx} = \frac{dy}{dw} \cdot \frac{dw}{dv} \cdot \frac{dv}{du} \cdot \frac{du}{dx}$

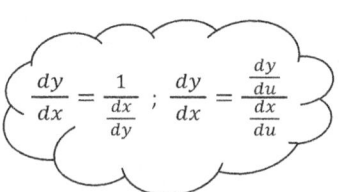

$\frac{dy}{dx} = \frac{1}{\frac{dx}{dy}}$; $\frac{dy}{dx} = \frac{\frac{dy}{du}}{\frac{dx}{du}}$

Note: $\boxed{y = \sin f(x) \Rightarrow y' = \cos f(x) \cdot f'(x)}$

$\because$ Let $u = f(x)$

Then, $y = \sin u$

$\therefore \frac{dy}{du} = \cos u$ and $\frac{du}{dx} = f'(x)$

$\therefore \frac{dy}{dx} = \frac{dy}{du} \cdot \frac{du}{dx} = \cos u \cdot f'(x)$

Therefore, $y' = \cos f(x) \cdot f'(x)$

Example Compute the derivative: (1) $y = \sin(x^3 + 1)$ (2) $y = \tan(\cos(\sin(x^3 + 1)))$

(1) $y = \sin(x^3 + 1)$

Let $u = x^3 + 1$ Then, $y = \sin u$

$\frac{du}{dx} = \frac{d}{dx}(x^3 + 1) = 3x^2$

$\frac{dy}{du} = \frac{d}{du}(\sin u) = \cos u$

$\therefore \frac{dy}{dx} = \frac{dy}{du} \cdot \frac{du}{dx} = \cos u \cdot 3x^2 = 3x^2 \cos u = 3x^2 \cos(x^3 + 1)$

(2) $y = \tan(\cos(\sin(x^3 + 1)))$

Let $u = x^3 + 1$; $v = \sin u = \sin(x^3 + 1)$; $w = \cos v = \cos(\sin(x^3 + 1))$;

    $y = \tan w = \tan(\cos(\sin(x^3 + 1)))$

Then, $\frac{du}{dx} = 3x^2$; $\frac{dv}{du} = \cos u$; $\frac{dw}{dv} = -\sin v$; $\frac{dy}{dw} = \sec^2 w$

$\therefore \frac{dy}{dx} = \frac{dy}{dw} \cdot \frac{dw}{dv} \cdot \frac{dv}{du} \cdot \frac{du}{dx} = \sec^2 w \cdot (-\sin v) \cdot \cos u \cdot 3x^2$

Example Evaluate $\dfrac{dy}{dx}$.

(1) $x = y\sqrt{1+y}$

$$\frac{dx}{dy} = (y)'(\sqrt{1+y}) + y(\sqrt{1+y})' = \sqrt{1+y} + y\left\{\frac{(1+y)'}{2\sqrt{1+y}}\right\}$$

$$= \sqrt{1+y} + \frac{y}{2\sqrt{1+y}} = \frac{2(1+y)+y}{2\sqrt{1+y}} = \frac{3y+2}{2\sqrt{1+y}}$$

$$\therefore \frac{dy}{dx} = \frac{1}{\frac{dx}{dy}} = \frac{1}{\frac{3y+2}{2\sqrt{1+y}}} = \frac{2\sqrt{1+y}}{3y+2}$$

(2) $x = t - 2, \ y = t^2$

$$x = t - 2 \Rightarrow \frac{dx}{dt} = 1 \ ; \ y = t^2 \Rightarrow \frac{dy}{dt} = 2t$$

$$\therefore \frac{dy}{dx} = \frac{dy}{dt} \cdot \frac{dt}{dx} = 2t \cdot \frac{1}{1} = 2t = 2(x+2)$$

(3) $x = 2\cos^3\theta, \ y = 2\sin^3\theta$

$$\frac{dx}{d\theta} = 2(3\cos^2\theta)(\cos\theta)' = 6\cos^2\theta(-\sin\theta) = -6\cos^2\theta\sin\theta$$

$$\frac{dy}{d\theta} = 2(3\sin^2\theta)(\sin\theta)' = 6\sin^2\theta(\cos\theta) = 6\sin^2\theta\cos\theta$$

$$\therefore \frac{dy}{dx} = \frac{dy}{d\theta} \cdot \frac{d\theta}{dx} = 6\sin^2\theta\cos\theta \cdot \frac{1}{-6\cos^2\theta\sin\theta} = \frac{-\sin\theta}{\cos\theta} = -\tan\theta$$

Example Find the slope of the tangent line at a point $(2, 1)$ on the graph of $y^3 + 2y = x^3$.

Assume that the given equation determines y as a function of x.

By the chain rule,

$$\frac{d}{dx}(y^3 + 2y) = \frac{d}{dx}(x^3) \ ; \ 3y^2\frac{dy}{dx} + 2\frac{dy}{dx} = 3x^2 \ ; \ \frac{dy}{dx}(3y^2 + 2) = 3x^2 \ ; \ \frac{dy}{dx} = \frac{3x^2}{3y^2+2}$$

$$\therefore \text{At } (2, 1), \ \frac{dy}{dx} = \frac{3x^2}{3y^2+2} = \frac{3\cdot2^2}{3\cdot1^2+2} = \frac{12}{5}$$

Therefore, the slope of the tangent line at $(2, 1)$ is $\dfrac{12}{5}$.

(3) Derivative of Inverse Function

When a function $y = f(x)$ is differentiable and its inverse function is $x = g(y)$,

$$\frac{dy}{dx} = \frac{1}{\frac{dx}{dy}} \left(\frac{dx}{dy} \neq 0\right) \; ; \; f'(x) = \frac{1}{g'(f(x))} = \frac{1}{g'(y)} \; (g'(y) \neq 0)$$

Example Find the derivative of the inverse function of $f(x) = x\sqrt{1+x} \; (x > 0)$.

$$f(x) = x\sqrt{1+x} \;\Rightarrow\; f'(x) = 1 \cdot \sqrt{1+x} + x\left(\sqrt{1+x}\right)' = \sqrt{1+x} + x \cdot \frac{(1+x)'}{2\sqrt{1+x}}$$

$$= \sqrt{1+x} + x \cdot \frac{1}{2\sqrt{1+x}} = \frac{2(1+x)+x}{2\sqrt{1+x}} = \frac{3x+2}{2\sqrt{1+x}}$$

Let g be the inverse function of f. Then, $g'(y) = \dfrac{1}{f'(g(y))} = \dfrac{2\sqrt{1+y}}{3y+2}$

3. Derivatives of the Logarithmic Functions

(1) Fundamental Properties of the Log function

1) $\boxed{\log ab = \log a + \log b, \text{ whenever } a > 0 \text{ and } b > 0}$

2) $\boxed{\log \dfrac{1}{b} = -\log b, \text{ whenever } b > 0}$

$\because$ Since $b \cdot \dfrac{1}{b} = 1, \; \log\left(b \cdot \dfrac{1}{b}\right) = \log 1 = 0$

Since $\log\left(b \cdot \dfrac{1}{b}\right) = \log b + \log\dfrac{1}{b}, \; \log b + \log\dfrac{1}{b} = 0$

$\therefore \; \log b = -\log\dfrac{1}{b}$

3) $\boxed{\log \dfrac{a}{b} = \log a - \log b, \text{ whenever } a > 0 \text{ and } b > 0}$

$\because$ Since $\dfrac{a}{b} = a \cdot \dfrac{1}{b}$,

$\log\dfrac{a}{b} = \log\left(a \cdot \dfrac{1}{b}\right) = \log a + \log\dfrac{1}{b} = \log a - \log b$

4) $\boxed{\log a^n = n \log a, \text{ whenever } a > 0 \text{ and } n \text{ is a positive integer}}$

$\because$ By mathematical induction,

i) $\log a^1 = \log a = 1 \cdot \log a \quad \therefore$ The statement is true if $n = 1$

ii) Assume it is true if $n = k$

iii) If $n = k+1, \; \log a^{k+1} = \log(a^k \cdot a) = \log a^k + \log a$ (by 1))

$$= k \log a + \log a \qquad\qquad (\text{ by ii) })$$

$$= (k+1)\log a \qquad\qquad (\text{ by the distributive law})$$

Therefore, the statement is true when $n = k + 1$.

Hence, by mathematical induction, the statement is true for any positive integer.

---

5) $\boxed{\log a^{\frac{1}{n}} = \left(\frac{1}{n}\right)\log a, \text{ whenever } a > 0 \text{ and } n \text{ is a positive integer}}$

$\because$ Since $\left(a^{\frac{1}{n}}\right)^n = a$, $\log\left(a^{\frac{1}{n}}\right)^n = \log a$

Since $\log\left(a^{\frac{1}{n}}\right)^n = n\log a^{\frac{1}{n}}$, (by 4))

$n\log a^{\frac{1}{n}} = \log a$ $\qquad \therefore \log a^{\frac{1}{n}} = \frac{1}{n}\log a$

---

6) $\boxed{\log a^{\frac{m}{n}} = \left(\frac{m}{n}\right)\log a, \text{ whenever } a > 0, \text{ and } n, m \text{ are positive integers}}$

$\because$ Since $a^{\frac{m}{n}} = \left(a^{\frac{1}{n}}\right)^m$, $\log a^{\frac{m}{n}} = \log\left(a^{\frac{1}{n}}\right)^m = m\log a^{\frac{1}{n}} = \left(m\cdot\frac{1}{n}\right)\log a = \left(\frac{m}{n}\right)\log a$

---

(2) Derivative of the Log Function

Note that $\lim_{x\to 0}(1 + x)^{\frac{1}{x}} = e$; $\lim_{x\to 0}\left(1 + \frac{1}{x}\right)^x = e$

For a log function defined for $x > 0$,

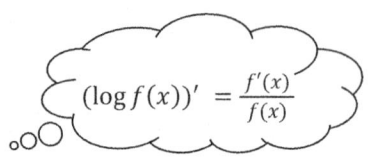

$(\log f(x))' = \dfrac{f'(x)}{f(x)}$

1) $\boxed{\text{The derivative of } y = \log x \text{ is } y' = \frac{1}{x}.}$

$\because\ y' = \lim_{h\to 0}\dfrac{\log(x+h)-\log x}{h} = \lim_{h\to 0}\dfrac{1}{h}\log\left(\dfrac{x+h}{x}\right) = \lim_{h\to 0}\dfrac{1}{h}\log\left(1 + \dfrac{h}{x}\right)$

$= \lim_{h\to 0}\log\left(1 + \dfrac{h}{x}\right)^{\frac{1}{h}} = \lim_{h\to 0}\log\left(1 + \dfrac{h}{x}\right)^{\frac{x}{h}\cdot\frac{1}{x}} = \lim_{h\to 0}\dfrac{1}{x}\log\left(1 + \dfrac{h}{x}\right)^{\frac{x}{h}}$

$= \dfrac{1}{x}\log\lim_{h\to 0}\left(1 + \dfrac{h}{x}\right)^{\frac{x}{h}} = \dfrac{1}{x}\log e = \dfrac{1}{x}$

For example, $y = x\log(x^2 + 1)$

$\Rightarrow y' = x'\log(x^2 + 1) + x\{\log(x^2 + 1)\}' = \log(x^2 + 1) + x\dfrac{(x^2+1)'}{x^2+1}$

$= \log(x^2 + 1) + x\dfrac{2x}{x^2+1} = \log(x^2 + 1) + \dfrac{2x^2}{x^2+1}$

---

2) $\boxed{\text{The derivative of } y = \log_a x \text{ is } y' = \dfrac{1}{x}\cdot\dfrac{1}{\log_e a} \quad (a > 0,\ a \neq 1)}$

$$\because y' = \lim_{h \to 0} \frac{\log_a(x+h) - \log_a x}{h} = \lim_{h \to 0} \frac{1}{h} \log_a\left(\frac{x+h}{x}\right) = \lim_{h \to 0} \frac{1}{h} \log_a\left(1 + \frac{h}{x}\right)$$

$$= \lim_{h \to 0} \log_a\left(1 + \frac{h}{x}\right)^{\frac{1}{h}} = \lim_{h \to 0} \log_a\left(1 + \frac{h}{x}\right)^{\frac{1}{h} \cdot \frac{x}{x}} = \log_a e^{\frac{1}{x}} = \frac{1}{x} \log_a e = \frac{1}{x} \cdot \frac{1}{\log_e a}$$

For example, $y = \log_2(x^2 + 1)$

$$\Rightarrow y' = \frac{1}{x^2+1} \cdot \frac{1}{\log 2} \cdot (x^2+1)' = \frac{2x}{(x^2+1)\log 2}$$

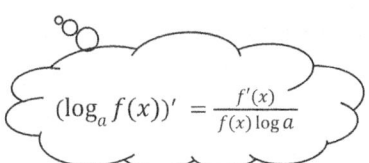

$$(\log_a f(x))' = \frac{f'(x)}{f(x)\log a}$$

Note : $\dfrac{d}{dx}\left(\log_a x\right)$ is the reciprocal of the logarithmic argument $\dfrac{1}{x}$ divided by the natural logarithm of the base $\ln a$. That is, $\dfrac{d}{dx}\left(\log_a x\right) = \dfrac{1}{x} \cdot \dfrac{1}{\ln a}$

4. Derivatives of the Exponential Function

Note that $\boxed{\lim_{x \to 0} \dfrac{a^x - 1}{x} = \log a \; ; \quad \lim_{x \to 0} \dfrac{e^x - 1}{x} = 1}$

$\because$ Let $a^x - 1 = z$ Then, $a^x = z + 1$ $\therefore \; x = \log_a(1+z)$

Since $z \to 0$ as $x \to 0$,

$$\lim_{x \to 0} \frac{a^x - 1}{x} = \lim_{z \to 0} \frac{z}{\log_a(1+z)} = \lim_{z \to 0} \frac{1}{\frac{1}{z}\log_a(1+z)} = \lim_{z \to 0} \frac{1}{\log_a(1+z)^{\frac{1}{z}}} = \frac{1}{\log_a e} = \log_e a = \log a$$

(1) $\boxed{\textbf{The derivative of } y = e^x \textbf{ is } y' = e^x}$

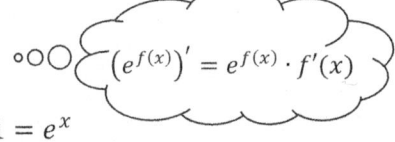

$$\left(e^{f(x)}\right)' = e^{f(x)} \cdot f'(x)$$

$$\because y' = \lim_{h \to 0} \frac{e^{x+h} - e^x}{h} = \lim_{h \to 0} \frac{e^x(e^h - 1)}{h} = e^x \lim_{h \to 0} \frac{e^h - 1}{h} = e^x \cdot 1 = e^x$$

For example, $y = e^{x^2} \sin x$

$$\Rightarrow y' = \left(e^{x^2}\sin x\right)' = \left(e^{x^2}\right)'\sin x + e^{x^2}(\sin x)' = e^{x^2}(x^2)'\sin x + e^{x^2}\cos x$$

$$= e^{x^2}(2x)\sin x + e^{x^2}\cos x = e^{x^2}(2x\sin x + \cos x)$$

(2) $\boxed{\textbf{The derivative of } y = a^x \; (a > 0, \; a \neq 1) \textbf{ is } y' = a^x \log a}$

$$\because y' = \lim_{h \to 0} \frac{a^{x+h} - a^x}{h} = \lim_{h \to 0} \frac{a^x(a^h - 1)}{h}$$

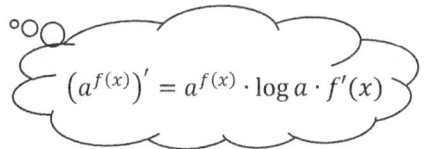

$$\left(a^{f(x)}\right)' = a^{f(x)} \cdot \log a \cdot f'(x)$$

$$= a^x \lim_{h \to 0} \frac{a^h - 1}{h} = a^x \log a$$

Note: $\boxed{y = a^{f(x)} \; \Rightarrow \; y' = a^{f(x)} \cdot \log a \cdot f'(x)}$

$\because$ Let $u = f(x)$ Then, $y = a^u$

$$\frac{dy}{du} = a^u \log a \; ; \quad \frac{du}{dx} = f'(x)$$

$$\therefore \; \frac{dy}{dx} = \frac{dy}{du} \cdot \frac{du}{dx} = a^u \cdot \log a \cdot f'(x)$$

For example, $y = a^{\sin x}$

$$\Rightarrow \quad y' = \left(a^{\sin x}\right)' = a^{\sin x} \cdot \log a \cdot (\sin x)' = (\log a) a^{\sin x} \cdot \cos x$$

5. Derivatives of the Absolute Function

Consider the derivative of the function $y = x^x \ (x > 0)$.

Taking the logarithm of both sides, we have $\log y = \log x^x$; $\log y = x \log x$

$$\therefore \quad \frac{d}{dx}(\log y) = \frac{d}{dx}(x \log x)$$

$$\therefore \quad \frac{1}{y} \cdot \frac{dy}{dx} = 1 \cdot \log x + x\frac{1}{x} \qquad \therefore \quad \frac{dy}{dx} = y(\log x + 1) \qquad \therefore \quad y' = x^x(\log x + 1)$$

Note: $\quad y = a^x \ \Rightarrow \ y' = a^x \log a$

$$y = x^x \ \not\Longrightarrow \ y' = x^x \log x$$

$$\boxed{y = \log|x| \ \Rightarrow \ y' = \frac{1}{x} \quad ; \quad y = \log|f(x)| \ \Rightarrow \ y' = \frac{1}{f(x)}f'(x)}$$

$\because$ i) When $x > 0$, $y = \log x \qquad \therefore \ y = \frac{1}{x}$

ii) When $x < 0$, $y = \log(-x) \qquad \therefore \ y = \frac{1}{-x}(-x)' = \frac{1}{-x}(-1) = \frac{1}{x}$

$\therefore \ y = \log|x| \ \Rightarrow \ y' = \frac{1}{x}$

Example Find the derivative of the following function.

(1) $y = x^{\sin x}$

Note that $x^{\sin x} > 0$

Taking log of both sides of $y = x^{\sin x}$, $\quad \log y = \log x^{\sin x} = \sin x \cdot \log x$

$$\therefore \quad \frac{d}{dx}(\log y) = \frac{d}{dx}(\sin x \cdot \log x)$$

$$\therefore \quad \frac{1}{y} \cdot \frac{dy}{dx} = (\sin x)' \cdot \log x + \sin x \cdot (\log x)'$$

$$\therefore \quad \frac{1}{y} \cdot \frac{dy}{dx} = \cos x \cdot \log x + \sin x \cdot \frac{1}{x}$$

$$\therefore \quad \frac{dy}{dx} = y\left(\cos x \cdot \log x + \sin x \cdot \frac{1}{x}\right) = x^{\sin x}\left(\cos x \cdot \log x + \frac{1}{x}\sin x\right)$$

(2) $y = \frac{x(x-1)^2}{(x+1)^3}$

Taking the absolute value of both sides, we have

$$|y| = \left|\frac{x(x-1)^2}{(x+1)^3}\right| = \frac{|x|\left|(x-1)^2\right|}{\left|(x+1)^3\right|}$$

Taking log of both sides, we have

$$\log|y| = \log\frac{|x|\left|(x-1)^2\right|}{\left|(x+1)^3\right|} = \log|x| + \log|(x-1)^2| - \log|(x+1)^3|$$

$$= \log|x| + 2\log|x-1| - 3\log|x+1|$$

$$\therefore \frac{d}{dx}(\log|y|) = \frac{d}{dx}(\log|x| + 2\log|x-1| - 3\log|x+1|)$$

$$\therefore \frac{1}{y}\cdot\frac{dy}{dx} = \frac{1}{x} + \frac{2}{x-1} - \frac{3}{x+1}$$

$$\therefore \frac{dy}{dx} = y\left(\frac{1}{x} + \frac{2}{x-1} - \frac{3}{x+1}\right) = y\left\{\frac{(x-1)(x+1)+2x(x+1)-3x(x-1)}{x(x-1)(x+1)}\right\} = y\left\{\frac{5x-1}{x(x-1)(x+1)}\right\}$$

$$= \frac{x(x-1)^2}{(x+1)^3}\left\{\frac{5x-1}{x(x-1)(x+1)}\right\} = \frac{(x-1)(5x-1)}{(x+1)^4}$$

3-4 Higher-Order Derivatives

When f is a function, its derivative f' is again a function.

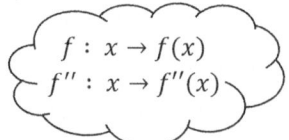

Thus, we can consider the derivative of f'.

If f' is differentiable at a point x, then the derivative of f' at x is denoted by $f''(x)$.

The number $f''(x)$ is called the *second derivative* of f at x (read "f double prime" or "f second").

For example, the function $f(x) = x^3 + 2x + 1$ has derivative $f'(x) = 3x^2 + 2$ and second derivative $f''(x) = 6x$.

If we continue the process of differentiating the function f, then the nth derivative of f, denoted by $f^{(n)}(x)$ or $\frac{d^n(f(x))}{dx^n}$, is defined in terms of the $(n-1)^{st}$ derivative by $\frac{d^n(f(x))}{dx^n} = \frac{d(f^{(n-1)}(x))}{dx}$.

For example, if $f(x) = x^3$, then $f^{(2)}(x) = f''(x) = f'(3x^2) = 6x$

$$\text{whereas } f^2(x) = \{f(x)\}^2 = (x^3)^2 = x^6$$

$f^{(n)}$ is the nth derivative
f^n is the nth power

Example Compute the first three derivatives of the function $y = \cos x$.

$$y' = (\cos x)' = \frac{d(\cos x)}{dx} = -\sin x$$

$$y'' = \frac{d^2(\cos x)}{dx^2} = \frac{d\{(\cos x)'\}}{dx} = \frac{d(-\sin x)}{dx} = -\cos x$$

$$y''' = \frac{d^3(\cos x)}{dx^3} = \frac{d^2(-\sin x)}{dx^2} = \frac{d(-\cos x)}{dx} = -(-\sin x) = \sin x$$

Note:

(1) The first derivative f' shows up in the graph of f as the slope.

(2) The second derivative f'' shows up as the rate of change of the slope.

    i) When f'' is positive, the slope is increasing.

    ii) When f'' is negative, the slope is decreasing.

For example, if $f(x) = \sin x$, then $f'(x) = \cos x$ and $f''(x) = -\sin x$

Therefore, the slope is decreasing when $\sin x > 0$

and increasing when $\sin x < 0$.

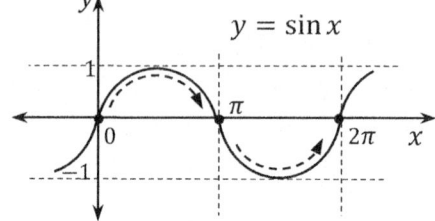

Exercises

#1 Find the difference quotient of each function f at the given interval.

(1) $f(x) = 2x + 3$ $[3, 6]$

(2) $f(x) = x^2 - 2x + 3$ $[1, 3]$

(3) $f(x) = \sqrt{x}$ $[4, 16]$

(4) $f(x) = x^3$ $[a, b]$

#2 When the graph of a function $y = f(x)$ is shown

as the Figure, let $g(x)$ be the inverse function of $f(x)$.

Find the difference quotient of $g(x)$ at the interval $[b, c]$.

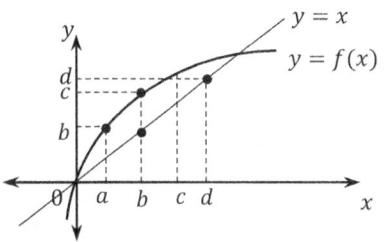

#3 Find the equation of the tangent to $y = x^2$ to the point $\left(x_0, f(x_0)\right)$.

#4 For a function $f(x) = ax^3 + bx + c$, the difference quotient of the function at the interval $[0, 1]$

is 6 and the derivative of f at $x = 1$ is 2. Find the value of $a - b$.

#5 Find the derivative of $f(x) = \dfrac{1}{x-2}$ at $x = 1$ and $x = 3$.

Show that the derivative does not exist at $x = 2$, where the function is discontinuous.

#6 Find the derivative of $f(x) = \dfrac{2x-3}{3x+4}$.

Examine the derivative at $x = -\dfrac{4}{3}$, where the function is discontinuous.

#7 Examine the differentiability of the function at $x = 0$.

(1) $f(x) = |x|(x - 1)$

(2) $f(x) = \begin{cases} x^2 \cos\dfrac{1}{x}, & x \neq 0 \\ 0, & x = 0 \end{cases}$

(3) $f(x) = x - [x]$, ([x]; the greatest integer in x)

#8 Examine each function is continuous at $x = 1$ but not differentiable.

(1) $f(x) = \sqrt{x - 1}$

(2) $f(x) = |x^2 - x|$

(3) $f(x) = \begin{cases} 2x^3 & , x \geq 0 \\ 4x - 2 & , x < 0 \end{cases}$

#9 Find the limit.

(1) $\lim\limits_{x \to 0} \dfrac{[\frac{3}{2}+x]-[\frac{3}{2}]}{x}$ ($[x]$; the greatest integer in x)

(2) When $f'(x) = 10$, $\lim\limits_{n \to \infty} n\left\{f\left(\dfrac{2}{n}\right) - f(0)\right\}$

(3) When $f(1) = 1$ and $f'(1) = 2$, i) $\lim\limits_{x \to 1} \dfrac{x^3 f(1) - f(x^2)}{x-1}$ ii) $\lim\limits_{x \to 1} \dfrac{xf(x)-1}{x^2-1}$

(4) When $f'(0) = 3$, $\lim\limits_{x \to 0} \dfrac{f(3x) - f(\sin x)}{x}$

(5) When $f(1) = \dfrac{1}{2}$ and $f'(1) = 4$, $\lim\limits_{x \to 1} \dfrac{f(x) - x^2 f(1)}{\sin(x-1)}$

(6) When $f(x) = \dfrac{1}{x}$, $\lim\limits_{x \to 1} \dfrac{f'(x)+1}{x-1}$

(7) When $f(x) = x^4 - 2x^3 + x + 3$, $\lim\limits_{x \to 0} \dfrac{f(1+x) - f(1-x)}{x}$

(8) For a function $f(x)$ such that i) $\lim\limits_{x \to 0} \dfrac{f(x)}{x} = \dfrac{1}{2}$ and ii) $f'(1) = 3$, $\lim\limits_{x \to 1} \dfrac{f(x) - f(1)}{f(x-1)}$

#10 Find the derivative of each of the following functions.

(1) $f(x) = \sqrt{2x+1}$

(2) $f(x) = \sqrt[3]{(x+1)(x^2+1)}$

(3) $f(x) = \dfrac{2x^3+x-1}{x^2}$

(4) $f(x) = \dfrac{1-x}{x^2+3}$

(5) $f(x) = (x+1)^3 (x^2-1)^2$

(6) $f(x) = \left(\dfrac{x}{x^2+1}\right)^3$

(7) $f(x) = \sqrt{\sqrt{x}+1}$

(8) $f(x) = (x+3)\sqrt{x-4}$

(9) $f(x) = \dfrac{3-2x}{\sqrt{x^2+1}}$

#11 Find the values of the constants a and b.

(1) For a function $f(x) = x^2 + ax + b$, $xf'(x) - 2f(x) + 2x^2 - 2 = 0$ is always true.
Find the values of the constants a and b.

(2) For a function $f(x) = x^4 + ax^2 + bx$, $\lim\limits_{x \to 2} \dfrac{f(x)-f(2)}{x-2} = 10$ and $\lim\limits_{x \to 1} \dfrac{f(x)-f(1)}{x^2-1} = -2$.
Find the values of the constants a and b.

(3) When a polynomial $x^5 - 5x + a$ is divided by $(x-b)^2$, there is no remainder. ($a > 0$, $b > 0$) Find the values of the constants a and b.

(4) For a graph of the function $f(x) = ax^2 + b$, the slope of the tangent line at $(1, 2)$ is -3.
Find the values of the constants a and b.

(5) For a polynomial $f(x)$ such that $\lim\limits_{h \to 0} \dfrac{f(4+h)+a}{h} = b$, θ is the angle between the tangent line

at $(4, -3)$ on the graph of $y = f(x)$ and x-axis. When $\tan \theta = \dfrac{1}{3}$, find the values of a and b.

(6) When a function $f(x) = \begin{cases} x^3 + ax^2 + bx , & x \geq 1 \\ 2x^2 + 1 , & x < 1 \end{cases}$ is differentiable at any real number x,

find the values of the constants a and b.

#12 Compute $\dfrac{dy}{dx}$.

(1) $y^2 = 3x$

(2) $y^3 = x^2$

(3) $\sqrt{x} + \sqrt{y} = 2$

(4) $x^3 + y^3 - 3xy = 0$

(5) $x = y\sqrt{1+y}$

(6) $x = 2t - 2,\ y = t^2$

#13 Find the value.

(1) When the graph of a function $y = f(x)$ is shown

as the Figure, the equation of the tangent line at $P(a, 5)$

on the graph is $y = 3x + b$. Find the value of the limit:

$$\lim_{n \to \infty} n \left\{ f\left(a + \frac{3}{n}\right) - f(a) \right\} \cdot f\left(a + \frac{2}{n}\right)$$

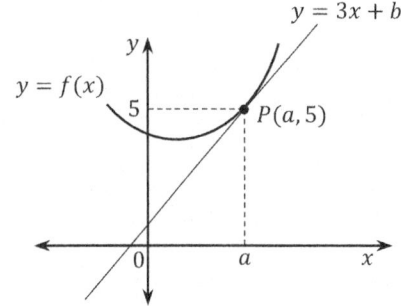

(2) When a polynomial $x^5 - 2x + 4$ is divided by $(x - 1)^2$, find the remainder.

(3) When a function $f(-x) = f(x)$ is differentiable at $x = 0$, find the value of $f'(0)$.

(4) For any real number x, a polynomial function $f(x)$ satisfies $f(-x) = f(x)$ and

$\lim\limits_{h \to 0} \dfrac{f(3-2h)-f(3+3h)}{h} = 10$. Find the value of $f'(-3)$.

(5) For a function $f(x)$ with degree 4 and leading coefficient 1, $f(x)$ satisfies:

(i) $f(x) = f(-x)$ for any real number x and

(ii) $\lim\limits_{h \to 1} \dfrac{f(2h)+1}{h-1} = 10$.

Find the value of $f'(-2)$.

(6) For a function defined by $f(x) = \begin{cases} 2\sin x + x^2 \cos\dfrac{1}{x^2}, & x \neq 0 \\ 0 , & x = 0 \end{cases}$, find the value of $f'(0)$.

(7) When $f(x) = (x + \sqrt{1 + x^2})^{10}$, find the value of $f'(1)f'(-1)$.

(8) When a function $f(x)$ is a polynomial for any x,

$f(x) = g'(x)$ and $\{f(x) + g(x)\}' = x^3 + 3x^2 + 4x + 5$. Find the value of $f(-1)$.

(9) For a composite function $F(x) = f(g(x))$ at which $f'(x) = \frac{1}{x^2+1}$, $g'(x) = \frac{1}{x^4+1}$, and $g(0) = 2$, find the value of $F'(0)$.

(10) Find the value of positive integer n such that $\lim_{x \to 1} \frac{x^{2n}+x^n-2}{x-1} = 15$.

(11) For a differentiable function $f(x)$ such that $f(x+y) = f(x) + f(y) + 2xy$, the tangent line at $x = 0$ is the x-axis. Find the value of $f'(1)$.

(12) A function $f(x)$ is differentiable and its inverse $f^{-1}(x)$ exists.
When $f(3f^{-1}(x) - 4x) = x$ for any real number x, find the value of $f^{-1}(5)$.

(13) A differentiable function $f(x)$ satisfies $f(x+y) = f(x)f(y)$ for any real numbers x and y. When $f'(0) = 4$ and $f(a) = 2$, find the value of $f'(3a)$. (a; constant)

(14) For two polynomials $f(x)$ and $g(x)$ such that

    (i) $f(0) = 1$, $f'(0) = -2$, $g(0) = 3$, and

    (ii) $\lim_{x \to 0} \frac{f(x)g(x)-3}{x} = 0$,

find the value of $g'(0)$.

(15) For any real number x, a polynomial $f(x)$ satisfies $(x^n - 2)f'(x) = f(x)$ and $f(4) = 5$, find the value of $f(10)$.

(16) For a polynomial $f(x)$ such that

    (i) $f(1) = 5$, and

    (ii) for any real number x, $(f \circ f)(x) = f(x)f'(x) + 3$,

find the value of $f(10)$.

(17) For positive integers a and b,

a function $f(x) = \lim_{n \to \infty} \frac{ax^{n+b}+2x-1}{x^n+1}$ $(x > 0)$ is differentiable at $x = 1$.

Find the value of $a - b$.

(18) For any positive integer n, a differentiable function $f(x)$ satisfies $f(nx) = n^3 f(x)$ and $f'(1) = 20$. Find the value of $f'\left(\frac{1}{2}\right)$.

#14 For a polynomial function $f(x)$ and positive integers m and n,

$\lim_{x \to \infty} \frac{f(x)}{x^m} = 1$, $\lim_{x \to \infty} \frac{f'(x)}{x^{m-1}} = a$, $\lim_{x \to 0} \frac{f(x)}{x^n} = b$, and $\lim_{x \to 0} \frac{f'(x)}{x^{n-1}} = 10$ (a, b; real numbers)

State whether the following statements are true or false.

(1) $m \geq n$ (2) $ab \geq 10$ (3) If $f(x)$ has degree of 3, then $am = bn$.

#15 For a function $f(x) = \begin{cases} 1 - x & , \ x < 0 \\ x^2 - 1 & , \ 0 \le x < 1 \\ \frac{2}{3}(x^3 - 1), & x \ge 1 \end{cases}$,

state whether the following statements are true or false.

(1) $f(x)$ is differentiable at $x = 1$.

(2) $|f(x)|$ is differentiable at $x = 0$.

(3) 2 is the minimum value of the positive integer k at which $x^k f(x)$ is differentiable at $x = 0$.

#16 Find the derivative of the function.

(1) $y = \sin \sqrt{1 - x^2}$

(2) $y = \cos(\sin x)$

(3) $y = \cos x^\circ$

(4) $y = (2x^2 + 1) \sin 2x$

(5) $y = \sin^3 x \cos 3x$

(6) $y = (\sec x + \tan x)^3$

(7) $y = \sin^2(2\pi x - a)$

(8) $y = \sec^3(2x + 5)$

(9) $y = 3e^x$

(10) $y = 2 \cdot 3^x$

(11) $y = xe^x$

(12) $y = e^x \sin x$

(13) $y = \frac{e^x}{e^x + 1}$

(14) $y = \frac{x}{3^x}$

(15) $y = (x^2 + e^x)^3$

(16) $y = \log x + x$

(17) $y = x \log x - x$

(18) $y = 3 \log_2 x - x^3$

(19) $y = e^x \log x$

(20) $y = \frac{\log x}{e^x}$

(21) $y = (\log_2 x)^3$

(22) $y = \log(\tan x + \sec x)$

(23) $y = \log(x + \sqrt{x^2 + 1})$

(24) $y = \log \frac{x-1}{x+1}$

(25) $y = a^{\sin x}$

(26) $y = \sin 2^x$

(27) $y = e^{x^2} \sin x$

(28) $y = \log_{10}(x^2 + 1)$

(29) $y = x \log(x^2 + 1)$

(30) $y = e^x \log(\sin x)$

(31) $y = e^{x^x}, \ x > 0$

#17 Compute $\frac{dy}{dx}$.

(1) $x = \sin y$

(2) $x = \cos y \ \ (0 < y < \pi)$

(3) $\sin x + \sin y = 1$

(4) $\sin x + \cos y = 1$

(5) $\cos(x + y) + \cos(x - y) = 1$

(6) $x = 2 \cos^3 \theta, \ y = 2 \sin^3 \theta$

(7) $x = 3 \cos t, \ y = 2 \sin t$

(8) $y^x = x^y \ \ (x > 0, \ y > 0)$

(9) $x = \frac{1 - t^2}{1 + t^2}, \ y = \frac{2t}{1 + t^2}$

#18 Calculate the second derivatives of each function.

(1) $y = \sqrt{x^2 + 1}$

(2) $y = \cos^3 x$

(3) $y = e^{x^2}$

(4) $y = \dfrac{x}{\log x}$

#19 For a function $f(x) = e^{ax} \sin x$,

(1) Find the values of $f'(\pi)$ and $f''(\pi)$.

(2) Find the value of the constant a such that $f''(x) - 2f'(x) + 2f(x) = 0$.

#20 Find the value.

(1) For a differentiable function $f(x)$, let $g(x) = \dfrac{x}{1-f(x)}$ and $g'(0) = \dfrac{1}{3}$.

Find the value of $f(0)$. $(f(x) \neq 1)$

(2) For a function $f(x) = x^2 e^{-x} \sin x$, find the value of $\lim\limits_{x \to 0} \dfrac{f'(x)}{x^2}$.

(3) For a function $f(x) = \lim\limits_{h \to 0} \dfrac{e^{x+h} - e^x}{\sqrt{x+h} - \sqrt{x}}$, find the value of $f'(1)$.

(4) For the graph of the curve $x^4 + y^3 + axy + b = 0$,

the value of $\dfrac{dy}{dx}$ at $(1, 0)$ on the graph is -2. Find the value of $a + b$. $(a, b;$ constants$)$

(5) For the function $f(x) = \dfrac{e^x \cos x}{1 + \sin x}$, find the value of $f'\left(\dfrac{\pi}{2}\right)$.

(6) For the function $f(x) = x^{\sqrt{x}}$ $(x > 0)$, find the range of x such that $f'(x) > 0$.

(7) For the function $f(x) = \log(\log x)$, find the value of $\lim\limits_{h \to 0} \dfrac{f(e+h) - f(e-h)}{h}$.

(8) Find the value of the limit: $\lim\limits_{x \to a} \dfrac{x^2 e^a - a^2 e^x}{x - a}$.

(9) For $A = \lim\limits_{x \to 0} \dfrac{2}{x} \log\left(\dfrac{e^x + e^{2x} + e^{3x} + \cdots\cdots + e^{nx}}{n}\right)$, express A as n.

(10) For the function $f(x) = \sin^2 \dfrac{x}{2}$, find the value of $\lim\limits_{x \to \pi} \dfrac{f'(x)}{x - \pi}$.

(11) When a differentiable function $f(x)$ satisfies $f(2x^2 + 3x) - f(5x) = (x^2 - 4x)^3 + 2x$ for any real number x, find the value of $f'(5)$.

(12) For a function $f(x) = x^{2\ln x}$, find the limit: $\lim\limits_{x \to 0} \dfrac{f(e+2x) - f(e-2x)}{x}$

(13) For a differentiable function $f(x)$, let $g(x)$ be the inverse function of $f(x)$.

When $\lim\limits_{h \to 0} \dfrac{g(2+2h) - g(2-h)}{h} = 15$ and $f(5) = 2$, find the value of $f'(5)$.

(14) For a differentiable function $f(x)$, let $g(x)$ be the inverse function of $f(x)$.

When $\lim\limits_{x \to 1} \dfrac{f(x) - 3}{x - 1} = 2$, find the value of $g'(3)$.

(15) For a function $f(x) = 4\sin\frac{x}{2}$ $\left(0 \le x \le \frac{\pi}{2}\right)$, let $g(x)$ be the inverse of $f(x)$.

Find the value of $g'(2)$.

(16) For a function $f(x) = \ln(\ln x)$ $(x > 1)$, let $g(x)$ be the inverse function of $f(x)$.

Find the value of $\dfrac{g'(0)}{g(0)}$.

(17) For a function $f(x) = \ln(e^x + 1)$, let $g(x)$ be the inverse function of $f(x)$.

Find the value of $\dfrac{1}{f'(a)} - \dfrac{1}{g'(a)}$. $(a > 0)$

(18) For the graph of a curve $x^3 + y^3 + 6xy = 1$,

find the slope of the tangent line at $(1, -2)$ on the graph.

(19) Find the slope of the curve $x = y^2 - 3y$ at the points where it crosses the y-axis.

(20) For any real number x, a function satisfies

i) $f(1) = 3$, $f'(1) = 2$ and ii) $\lim\limits_{x \to 1} \dfrac{f'(f(x)) - 1}{x - 1} = 4$. Find the value of $f''(3)$.

(21) For a function $f(x) = \dfrac{ax+1}{x^2+bx}$ defined on any non-zero real number x, the sum and product

of two roots of $f'(x) = 0$ are -1 and $\dfrac{1}{2}$, respectively. When the equation $f'(x) + c = 0$ has

a double root, find the value of c. $(ab(\ne 0)$; real number$)$

(22) The graph of a function $f(x) = \ln x$ is shown as the Figure.

For a point P on the graph and a point Q on the x-axis,

the line segment $\overline{PQ}$ and the x-axis are perpendicular.

For a point $A(1, 0)$, let $\overline{PQ} = a$ and $\overline{PA} = f(a)$.

Find the value of $f'(\ln 3)$.

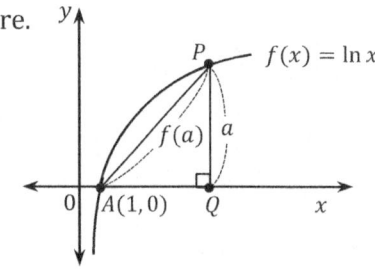

#21 Find the value of $a + b$.

(1) When a function $f(x) = \begin{cases} a\cos\frac{\pi}{2}x + 1, & x > 1 \\ e^{x-1} + b, & x \le 1 \end{cases}$ is differentiable at $x = 1$,

find the value of $a + b$.

(2) When a function $f(x) = \begin{cases} \log_2(x-1), & x \ge 2 \\ a\tan\pi(x-1) + b, & x < 2 \end{cases}$ is differentiable at $x = 2$,

find the value of $a + b$.

(3) When a function $f(x) = \begin{cases} a\ln(3x-2) + 2, & x > 1 \\ e^{x^2-1} + b, & x \le 1 \end{cases}$ is differentiable at $x = 1$,

find the value of $a + b$.

(4) When the limit is $\lim\limits_{x \to a} \dfrac{b \log x}{x^2 - a^2} = 1$, find the value of $a + b$.

(5) When a function $f(x) = \dfrac{ax+b}{x^2+1}$ satisfies

i) $\lim\limits_{x \to 3} \dfrac{f(x)-f(3)}{x-3} = -\dfrac{11}{50}$ and ii) $\lim\limits_{x \to 1} \dfrac{f(x)-f(1)}{x^2-1} = 0$, find the value of $a + b$.

Chapter 4. Applications of the Derivative

4-1 The Normal to a Curve

1. Equations of Tangent and Normal Lines

Consider the geometrical meaning of the derivative.

For the graph of $y = f(x)$,

take a point $P = P(x_0, y_0)$ and

a nearby point $Q = Q(x_0 + h, f(x_0 + h))$ on the graph.

The secant line through P and Q has its slope m_{sec}

equal to the difference quotient at x_0; i.e.,

$$m_{sec} = \frac{f(x_0 + h) - f(x_0)}{h}$$

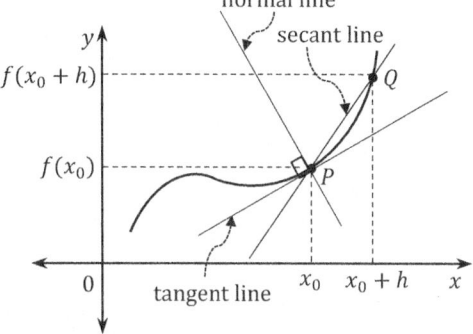

As Q slides along the curve to coincide with P, the secant line approaches the tangent line.

That is, the slope m of the tangent line at x_0 is $m = \lim\limits_{h \to 0} \frac{f(x_0 + h) - f(x_0)}{h} = f'(x_0)$

Therefore, if the function $y = f(x)$ has a derivative $f'(x_0)$ at x_0, then the curve has a tangent at

$P(x_0, y_0)$ where slope is $m = f'(x_0)$.

Hence the equation of the tangent line at (x_0, y_0) is $\boxed{y - y_0 = f'(x_0)(x - x_0)}$

Note: (1) A line perpendicular to a horizontal line is vertical.

      (2) A line perpendicular to a vertical line is horizontal.

      (3) If a line is neither horizontal nor vertical, then it has slope $m(\neq 0)$, and

         the line perpendicular to it has slope $-\dfrac{1}{m}$.

> The slope of the normal line is the negative reciprocal of the slope of the tangent line.

Normal Line

If a function $y = f(x)$ is differentiable at x_0, and therefore, has a unique tangent line to its

graph at (x_0, y_0), then the normal line to the function at (x_0, y_0) is the line passing through the

point (x_0, y_0) and perpendicular to the tangent line there.

Thus, the equation of the normal line at (x_0, y_0) is $\boxed{y - y_0 = -\dfrac{1}{f'(x_0)}(x - x_0), \ f'(x_0) \neq 0}$

or $f'(x_0)(y - y_0) + (x - x_0) = 0$

Example   Write the equation of the tangent and normal lines to $f(x) = -2x^2 + 3x + 7$ at $x = -1$.

Note that the point of tangency lies on the graph of $f(x)$.

When $x = -1$, the corresponding y-coordinate is $f(-1) = -2(-1)^2 + 3(-1) + 7 = 2$.

$\therefore$ The point of tangency is $(-1, 2)$.

Note that the slope of the tangent line at the point $(-1, 2)$ is $f'(-1)$.

Since $f'(x) = -4x + 3$, $f'(-1) = -4(-1) + 3 = 7$

Thus, the equation of the tangent line to $f(x)$ at $x = -1$ is $y - f(-1) = f'(-1)(x - (-1))$.

That is, $y - 2 = 7(x + 1)$; $y = 7x + 9$

The equation of the normal line at $(-1, 2)$ is $y - 2 = -\dfrac{1}{7}(x + 1)$; $y = -\dfrac{1}{7}x + \dfrac{13}{7}$

4-2 The Mean Value Theorem

Let us remind the intermediate-value theorem applies to any function f continuous on a closed interval $[a, b]$.

Now, we add the condition that f is differentiable on the open interval (a, b).

The differentiability allows us to find a point x in $[a, b]$ such that $f(x)$ is an extreme value of f on $[a, b]$.

1. Maxima and Minima

For a given function f in its domain D,

(1) $f(c)$ is the *maximum value* of f on D if $f(c) \geq f(x)$ for all x in D.

(2) $f(c)$ is the *minimum value* of f on D if $f(c) \leq f(x)$ for all x in D.

(3) $f(c)$ is an *extreme value* of f on D if it is either the maximum value or the minimum value.

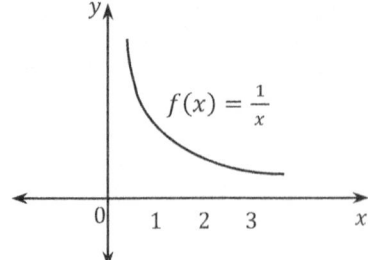

On $D = (0, \infty)$, f has no maximum or minimum.

On $D = [1, 3]$, maximum value $= 1$, minimum value $= \dfrac{1}{3}$

On $D = (1, 3]$ No maximum value, minimum value $= \dfrac{1}{3}$

2. Maximum and Minimum Existence Theorem

If f is a continuous on a closed interval $[a, b]$, then f attains both a maximum value and a minimum value there.

(1) If c is a point at which $f'(c) = 0$, then c is called a *stationary point*.

(2) If c is a point in an open interval where $f'(c)$ does not exist, then c is called a *singular point*. In this case, the graph of f has a sharp corner, a vertical tangent, takes a jump, or wiggles very badly.

(3) Any point in the domain of a function f of one of these (end points, stationary point, and singular point) is called a *critical point* of f.

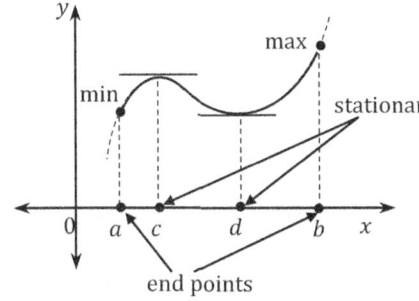
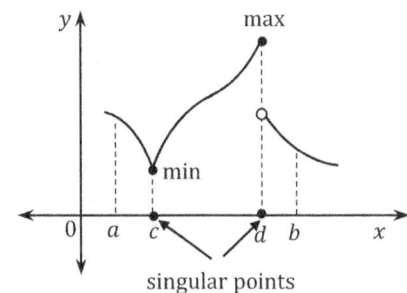

Critical Point Theorem

Let f be defined on an interval containing the point c.

If $f(c)$ is an extreme value, then c must be a critical point.

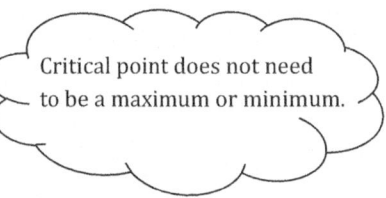

Critical point does not need to be a maximum or minimum.

Example Find the extreme values of the function $y = x^3$ on $[-1, 2]$.

The extreme values must occur at $x = -1$, at $x = 2$, or at a stationary point of the function.

$f'(x) = 3x^2 = 0 \Rightarrow x = 0$

$f(-1) = -1$

$f(2) = 8$

$f(0) = 0$

∴ The maximum on $[-1, 2]$ occurs at $x = 2$ and the minimum occurs at $x = -1$.

No extreme value occurs at the other critical point.

3. The Extreme Value Test

Let f be continuous on a closed interval $[a, b]$ and differentiable on an open interval (a, b).

If $a < c < b$ and $f(c)$ is either the maximum or minimum of f on $[a, b]$, then $f'(c) = 0$.

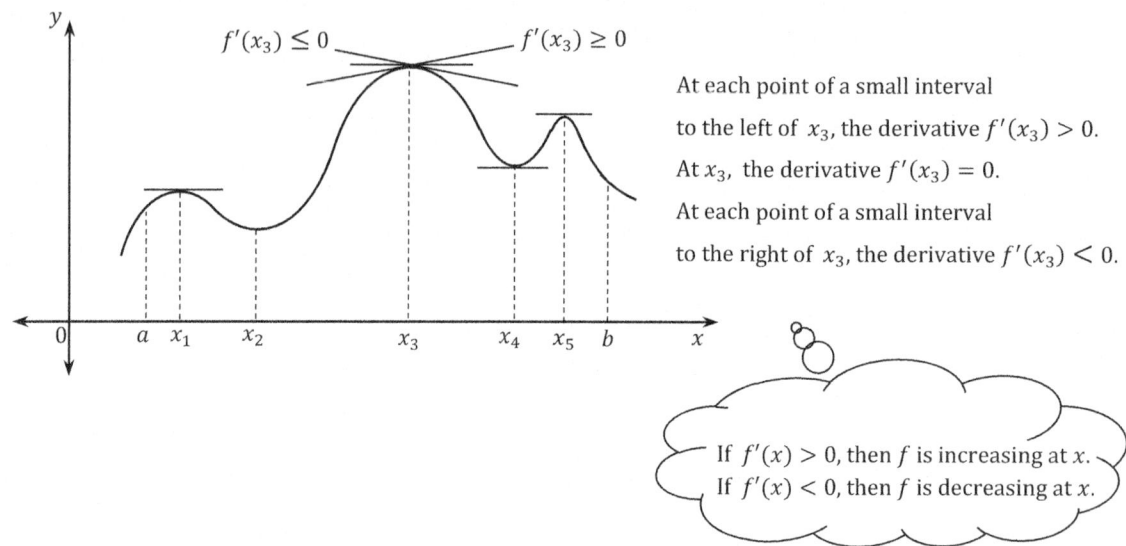

At each point of a small interval
to the left of x_3, the derivative $f'(x_3) > 0$.
At x_3, the derivative $f'(x_3) = 0$.
At each point of a small interval
to the right of x_3, the derivative $f'(x_3) < 0$.

If $f'(x) > 0$, then f is increasing at x.
If $f'(x) < 0$, then f is decreasing at x.

The extreme value test indicates that:

if a smooth curve crosses the x-axis at $x = a$ and again at $x = b$, then there should be at least one critical point between a and b.

4. Rolle's Theorem

If f is continuous on a closed interval $[a, b]$, differentiable on an open interval (a, b), and if $f(a) = f(b) = 0$, then there is at least one number c in (a, b) such that $f'(c) = 0$.

Note: If f and g are continuous on $[a, b]$ and differentiable on (a, b), and if $f(a) = g(a)$ and $f(b) = g(b)$, then there is at least one number c in (a, b) such that $f'(c) = g'(c)$.

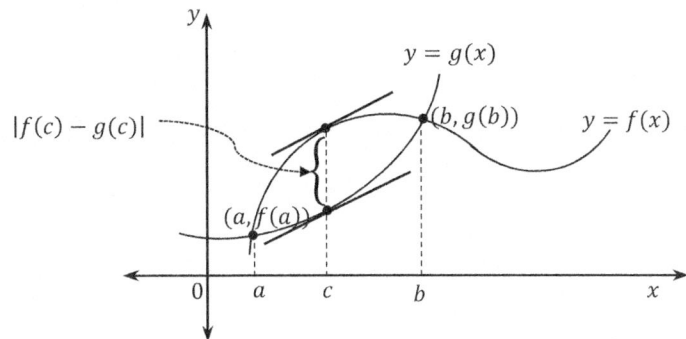

Note: If f is not differentiable in an open interval (a, b), then the Rolle's Theorem is not hold.

For example, let $f(x) = |x - 1|$.

Then, $f(x)$ is continuous on $[0, 2]$ and $f(0) = f(2) = 1$.

But, there is no c in $(0, 2)$ such that $f'(c) = 0$.

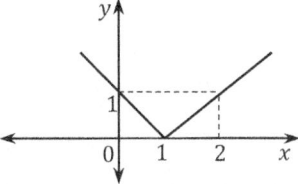

Example   For $f(x) = 12x - 3x^2$, find the value of c in $(0, 4)$ such that $f'(c) = 0$.

Since $f(x)$ is continuous in the closed interval $[0, 4]$ and differentiable in the open interval $(0, 4)$, and $f(0) = f(4) = 0$, by the Rolle's Theorem, there is c in $(0, 4)$ such that $f'(c) = 0$.

Since $f'(x) = 12 - 6x$, $f'(c) = 12 - 6c = 0$ $\therefore c = 2$

5. Mean Value Theorem (For Derivatives)

If f is continuous on a closed interval $[a, b]$ and differentiable on an open interval (a, b), then there is at least one number c in (a, b) where $\dfrac{f(b) - f(a)}{b - a} = f'(c)$

or, equivalently, where $f(b) - f(a) = f'(c)(b - a)$.

Note: The mean value theorem means that for two points P and Q on the graph given by $y = f(x)$, there is a point c between them where the slope of the tangent line at $(c, f(c))$ is parallel to the secant line connecting P and Q.

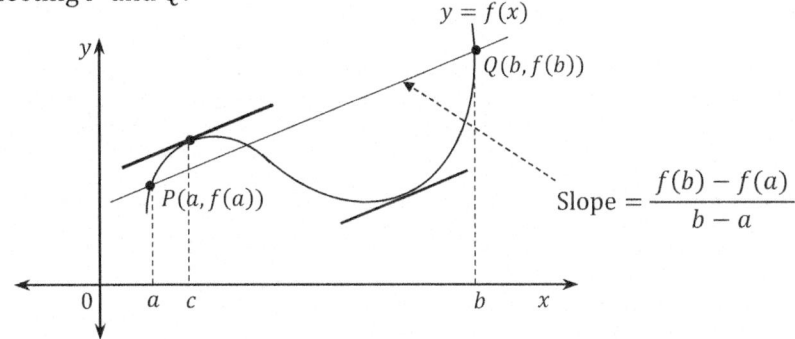

4-3 Monotonicity and Concavity

1. Monotonicity

Let f be a function defined on an interval (open, closed, or neither).

(1) If for any pair of numbers a and b in the interval, $a < b \Rightarrow f(a) < f(b)$,

then f is *increasing* on the interval.

(2) If for any pair of numbers a and b in the interval, $a < b \Rightarrow f(a) > f(b)$,

then f is *decreasing* on the interval.

(3) If it is either increasing on the interval or decreasing on the interval,

then f is *strictly monotonic* on the interval.

2. The First Derivative and Monotonicity

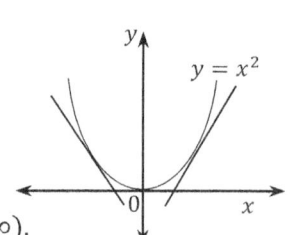

Consider a function $f(x) = x^2$.

Since $f'(x) = 2x$, $f'(x) > 0$ in $(0, \infty)$.

Form the graph of the function, $f(x)$ is increasing on the interval $(0, \infty)$.

Similarly, $f'(x) < 0$ in $(-\infty, 0)$ and $f(x)$ is decreasing on the interval $(-\infty, 0)$.

The first derivative $f'(x)$ gives us the slope of the tangent line to the graph of f at the point x.

Monotonicity Theorem

Let f be continuous on $[a, b]$ and differentiable on (a, b).

(1) If $f'(x) > 0$ for all x in (a, b), then f is increasing on $[a, b]$.

(2) If $f'(x) < 0$ for all x in (a, b), then f is decreasing on $[a, b]$.

> The sign of the derivative indicates where the function is increasing or decreasing.

> If $f'(x) = 0$,
> then $f(x)$ is a constant function.

Consider the functions $f(x) = x^3$ and $g(x) = -x^3$.

$f(x) = x^3 \Rightarrow f'(x) = 3x^2 \geq 0$ and $f'(0) = 0$

$g(x) = -x^3 \Rightarrow g'(x) = -3x^2 \leq 0$ and $g'(0) = 0$

$\therefore f(x) = x^3$ is increasing function but $f'(x) \geq 0$; i.e., $f'(x) > 0$ or $f'(x) = 0$

$g(x) = -x^3$ is decreasing function but $g'(x) \leq 0$; i.e., $g'(x) < 0$ or $g'(x) = 0$

Therefore, we have the following results:

For a differentiable function $f(x)$ on an interval,

(1) If $f(x)$ is increasing, then $f'(x) \geq 0$ in the interval.

(2) If $f(x)$ is decreasing, then $f'(x) \leq 0$ in the interval.

Note:

(1) <u>Concave Upward:</u>

For each point of a small interval to the left of x such that $f'(x) = 0$, $f'(x) < 0$.

For each point of a small interval to the right of x such that $f'(x) = 0$, $f'(x) > 0$.

(2) <u>Concave Downward:</u>

For each point of a small interval to the left of x such that $f'(x) = 0$, $f'(x) > 0$.

For each point of a small interval to the right of x such that $f'(x) = 0$, $f'(x) < 0$.

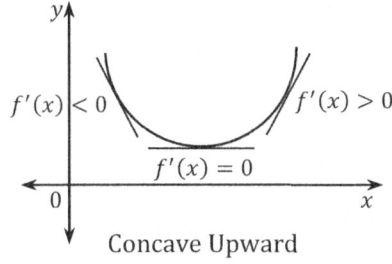

Concave Upward

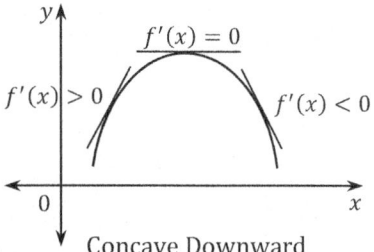

Concave Downward

3. The Second Derivative and Concavity

The first derivative f' shows up in the graph of f as the slope; the second derivative f'' shows up the rate of change of the slope.

(1) Concavity Theorem

A differentiable function f on an open interval is said to be *concave upward* if every tangent to the graph of f intersects the graph only at the point of tangency and otherwise lies entirely <u>below</u> the graph of f.

A differentiable function f on an open interval is said to be *concave downward* if every tangent to the graph of f intersects the graph only at the point of tangency and otherwise lies entirely <u>above</u> the graph of f.

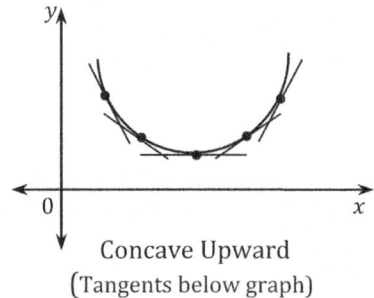

Concave Upward
(Tangents below graph)

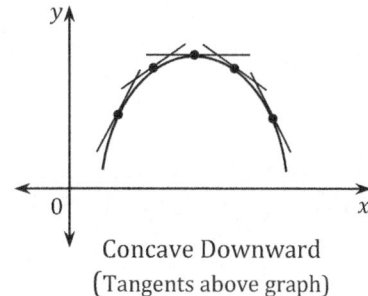

Concave Downward
(Tangents above graph)

Note: For a differentiable function f on an open interval,

      (1) if f' is monotone increasing ($f' > 0$), then f is concave upward and

      (2) if f' is monotone decreasing ($f' < 0$), then f is concave downward.

Note that the second derivative of f is the first derivative of f'.

Thus we have the following results:

(1) If f'' is positive then f' is increasing.

(2) If f'' is negative then f' is decreasing.

Concavity Theorem

Let f be a twice differentiable function on an open interval.

If $f''(x) > 0$ for all x, then f is concave upward;

If $f''(x) < 0$ for all x, then f is concave downward.

For example, $f(x) = 3x^2 - \cos x$

$\Rightarrow f'(x) = 6x + \sin x \ ; \ f''(x) = 6 + \cos x$

Since $f''(x) > 0$ for all x, f' is increasing and f is concave upward.

Example   Where is $f(x) = 5x^3 + 2x^2 - 3x$ increasing, decreasing, concave upward and concave downward?

$f'(x) = 15x^2 + 4x - 3 = (3x - 1)(5x + 3)$

$f''(x) = 30x + 4 = 2(15x + 2)$

$f'(x) > 0 \iff x > \frac{1}{3}$ and $x < -\frac{3}{5}$; $f'(x) < 0 \iff -\frac{3}{5} < x < \frac{1}{3}$

$\therefore \ f$ is increasing on $\left(-\infty, -\frac{3}{5}\right)$ and $\left(\frac{1}{3}, \infty\right)$ and f is decreasing on $\left(-\frac{3}{5}, \frac{1}{3}\right)$

When $x > -\frac{2}{15}$, $f''(x) > 0$; When $x < -\frac{2}{15}$, $f''(x) < 0$

$\therefore \ f$ is concave up on $\left(-\frac{2}{15}, \infty\right)$ and concave down on $\left(-\infty, -\frac{2}{15}\right)$.

(2) Inflection Points

Let f be continuous at c.

The point $(a, f(a))$ at which f is concave up on one side of c and concave down on the other side is called an *inflection point* of the graph of f. If f is differentiable for all values of x, then $f''(x) = 0$ or $f''(x)$ does not exist at its inflection points.

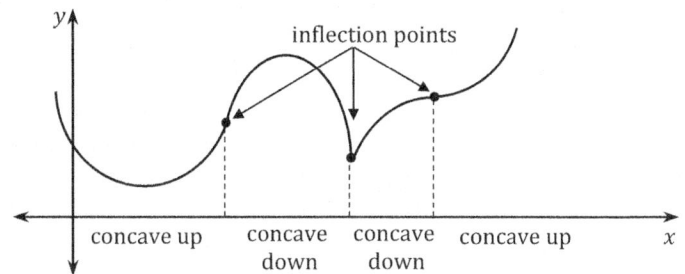

Example   Find all inflection points of the graph of (1) $f(x) = \frac{1}{3}x^3 - 4x$ (2) $f(x) = x^{\frac{1}{3}} + 4$.

(1) $f(x) = \frac{1}{3}x^3 - 4x \Rightarrow f'(x) = x^2 - 4$; $f''(x) = 2x$

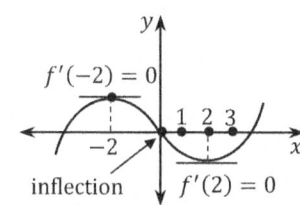

$\quad f''(x) = 2x = 0 \iff x = 0$

$\quad \therefore$ The inflection point is $(0, 0)$.

$\quad$ In fact, $f''(x) < 0$ for $x < 0$ and $f''(x) > 0$ for $x > 0$

(2) $f(x) = x^{\frac{1}{3}} + 4 \Rightarrow f'(x) = \frac{1}{3}x^{-\frac{2}{3}}$; $f''(x) = -\frac{2}{9}x^{-\frac{5}{3}} = -\frac{2}{9x^{\frac{5}{3}}}$

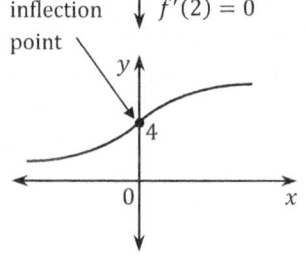

$\quad \therefore$ $f''(x)$ fails to exist at $x = 0$

$\quad$ Note that when $x < 0$, $f''(x) > 0$ and when $x > 0$, $f''(x) < 0$

$\quad \therefore$ The inflection point is $(0, 4)$ is

(3) Tests for Local Extreme Values

1) First Derivative Test for Local Extreme Values

Let f be continuous on an open interval (a, b) that contains a critical point c.

① If $f'(x) > 0$ for all x in (a, c) and $f'(x) < 0$ for all x in (c, b), then $f(c)$ is a *local maximum value* of f.

② If $f'(x) < 0$ for all x in (a, c) and $f'(x) > 0$ for all x in (c, b), then $f(c)$ is a *local minimum value* of f.

③ If $f'(x)$ has the same sign on both sides of c, then $f(c)$ is not a local extreme value of f.

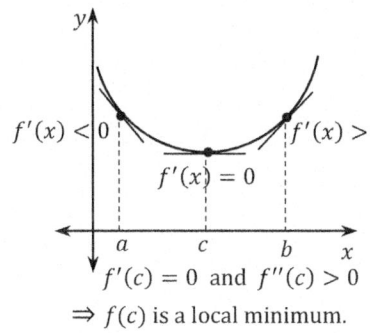

$f'(c) = 0$ and $f''(c) > 0$
$\Rightarrow f(c)$ is a local minimum.

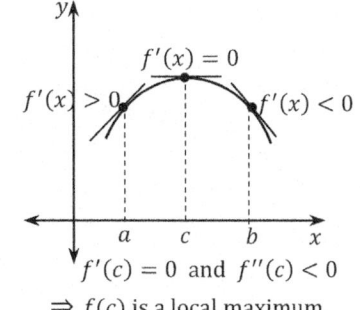

$f'(c) = 0$ and $f''(c) < 0$
$\Rightarrow f(c)$ is a local maximum.

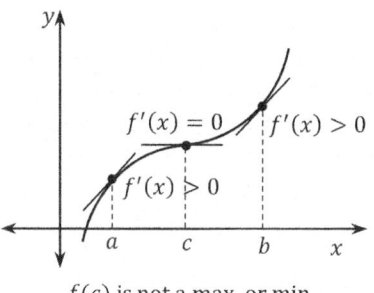

$f(c)$ is not a max. or min.

Example   Find the local extreme values of $f(x) = 2x^3 + 3x^2 - 12x - 4$ on $(-\infty, \infty)$.

Since $f'(x) = 6x^2 + 6x - 12 = 6(x^2 + x - 2) = 6(x + 2)(x - 1)$,

the critical points of f are -2 and 1.

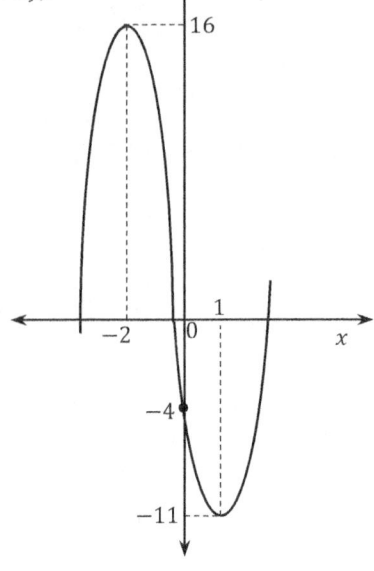

x	⋯⋯	-2		1	⋯⋯
$f'(x)$	+	0	−	0	+
$f(x)$	↗	local max.	↘	local min.	↗

$f'(x) > 0$ on $x > 1$ and $x < -2$; i.e., $(1, \infty)$ and $(-\infty, -2)$

$f'(x) < 0$ on $-2 < x < 1$; i.e., $(-2, 1)$

$\therefore$ $f(-2) = 16$ is a local maximum value and

$f(1) = -11$ is a local minimum value.

2) Second Derivative Test for Local Extreme Values

Let f' and f'' exist at all points in an open interval (a, b) containing c, and $f'(c) = 0$.

① If $f''(c) < 0$, then $f(c)$ is a local maximum value of f.

② If $f''(c) > 0$, then $f(c)$ is a local minimum value of f.

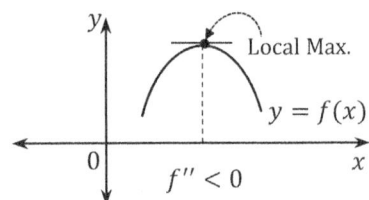   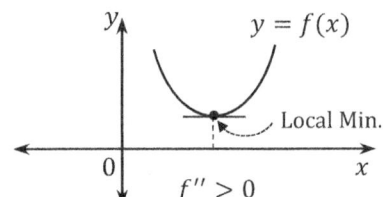

Example   For $f(x) = x^2 - 2x + 3$, use the second derivative test to identify local extreme values.

$f'(x) = 2x - 2 = 2(x - 1)$; $f''(x) = 2$ $\therefore$ $f'(1) = 0$ and $f''(1) > 0$

Therefore, by the second derivative test, $f(1)$ is a local minimum value.

Graphs of Polynomials

(1) $f(x) = ax + b$ $(a \neq 0)$

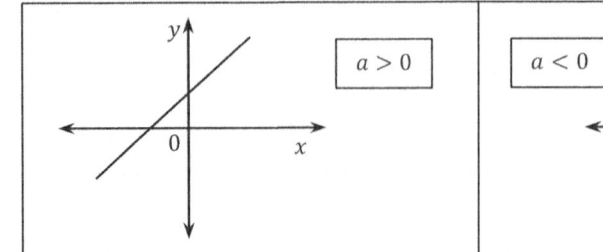   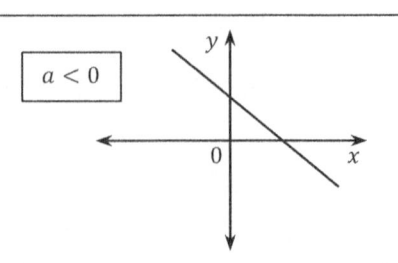

(2) $f(x) = ax^2 + bx + c \quad (a \neq 0)$

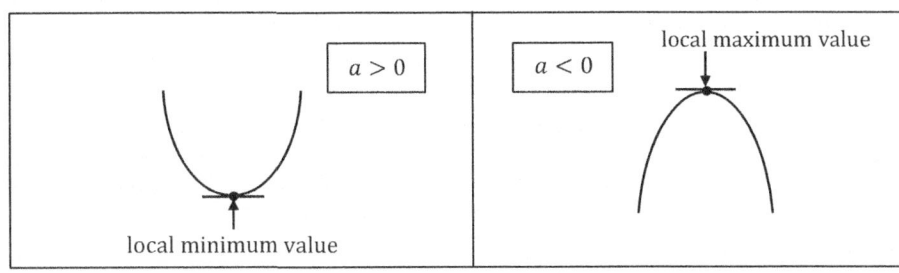

(3) $f(x) = ax^3 + bx^2 + cx + d \quad (a \neq 0)$

 i) When $f(x) = 0$ has three different real number solutions, $\alpha, \beta,$ and γ;

   (local maximum value) $\times$ (local minimum value) < 0

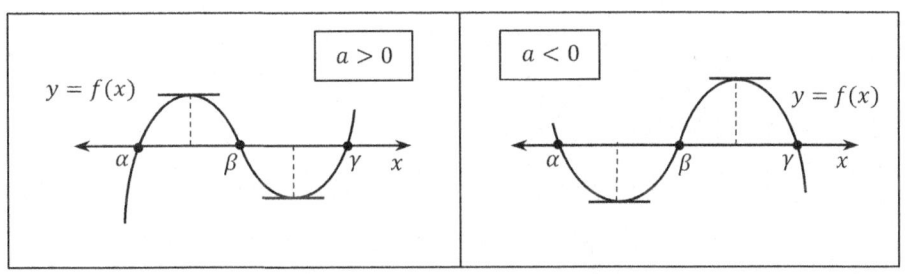

 $f'(x) = 0$ has two different real number solutions.

 $\Leftrightarrow\;$ $f(x)$ has local extreme values.

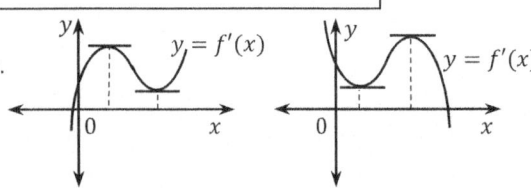

 ii) When $f(x) = 0$ has a double root, and a real number root;

   (local maximum value) $\times$ (local minimum value) $= 0$

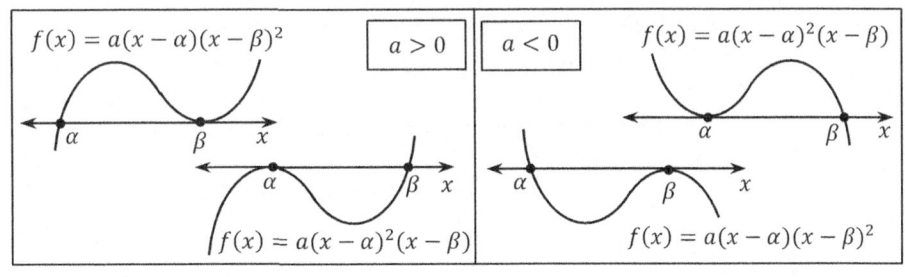

 $f'(x) = 0$ has a double root.

 $\Leftrightarrow\;$ $f(x)$ does not have local extreme values.

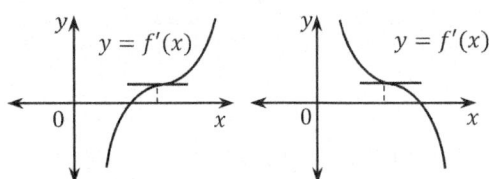

iii) When $f(x) = 0$ has a real number solution and two complex number solutions,

(local maximum value) $\times$ (local minimum value) > 0

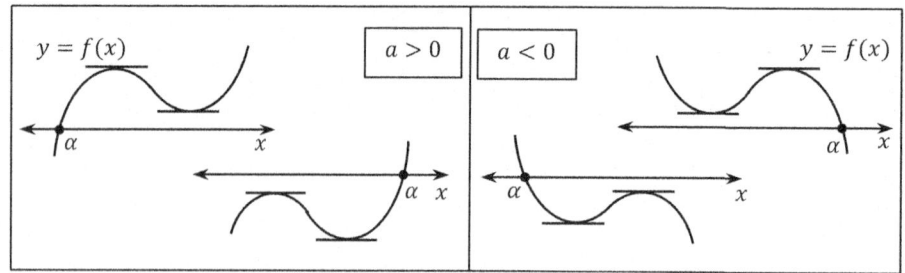

$f'(x) = 0$ has no real number solutions.

$\Leftrightarrow$ $f(x)$ does not have local extreme values.

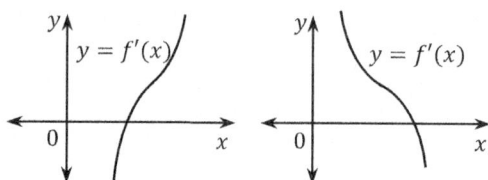

(4) $f(x) = ax^4 + bx^3 + cx^2 + dx + e$ $(a \neq 0)$

i) When $f'(x) = 0$ has three real number solutions, α, β, and γ ;

$f'(x) = a(x - \alpha)(x - \beta)(x - \gamma)$, $\alpha < \beta < \gamma$,

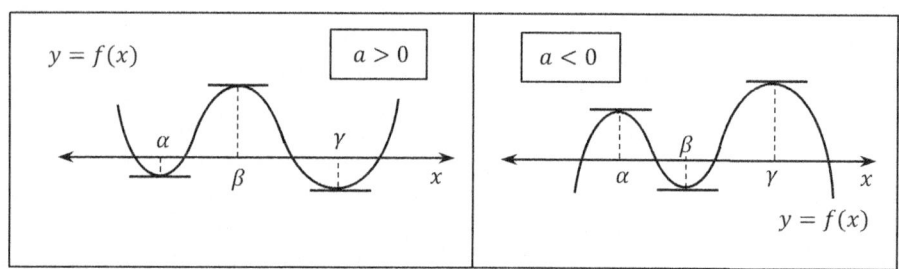

ii) When $f'(x) = 0$ has a double root, α, and a real number solution β;

$f'(x) = a(x - \alpha)^2(x - \beta)$, $\alpha < \beta$,

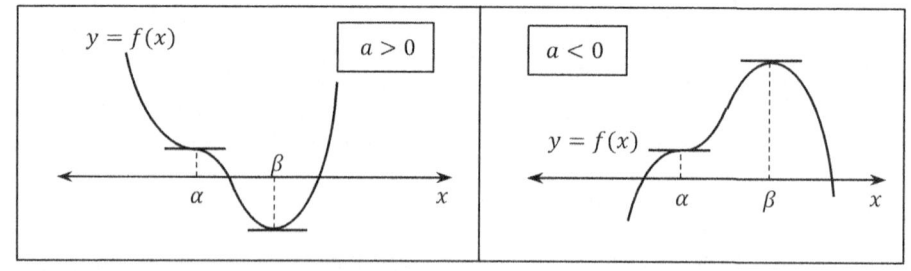

$f(x)$ has no local extreme values. $\Leftrightarrow$ $f'(x) = 0$ has a double root and only one real number solution.

iii) When $f'(x) = a(x - \alpha)^3$,

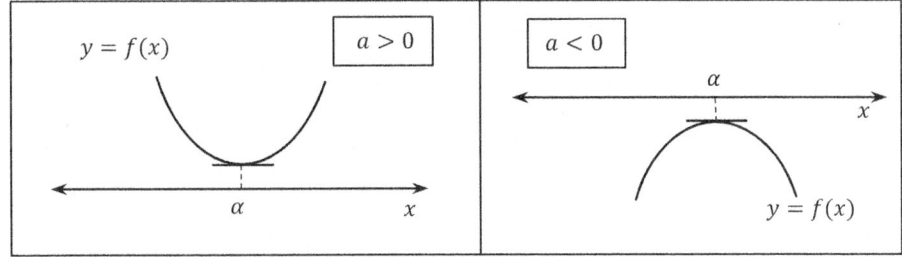

iv) When $f'(x) = 0$ has only one real number solution α,

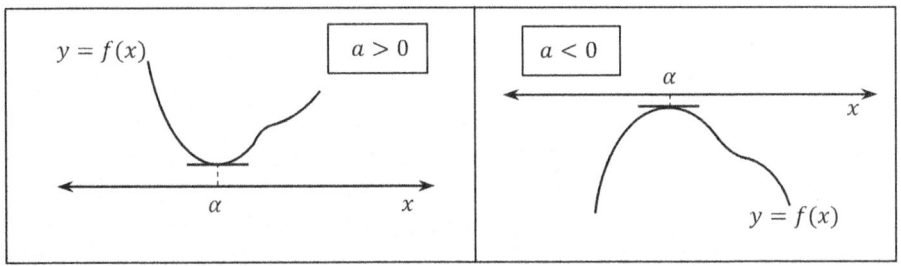

$f(x)$ has no local extreme values. $\iff$ $f'(x) = 0$ has only one real number solution

and two complex number solutions.

4-4 L'Hôpital's Rule

1. L'Hôpital's Rule for Forms of Type $\dfrac{0}{0}$

If the denominator has a non-zero limit, our basic quotient rule on limits is

$$\lim_{x \to a} \frac{f(x)}{g(x)} = \frac{\lim_{x \to a} f(x)}{\lim_{x \to a} g(x)}, \quad \lim_{x \to a} g(x) \neq 0.$$

But, in many limits of ratios, the denominator has a zero limit.

For example,

$$\lim_{x \to 0} \frac{1}{x}, \quad \lim_{x \to 0} \frac{\cos x}{x} \quad \Rightarrow \text{The limit does not exist.}$$

$\quad$ ($\because$ The denominator goes to zero, while the numerator does not.)

$$\lim_{x \to 0} \frac{\sin x}{x}, \quad \lim_{x \to 2} \frac{x^2 - 5x + 6}{x^2 - 4}, \quad \lim_{x \to a} \frac{f(x) - f(a)}{x - a}$$

$\Rightarrow$ Both numerator and denominator have 0 as their limits.

So that the quotient rule for limits is not available for their evaluation.

These are called "limits of the form $\frac{0}{0}$".

They can be evaluated by L'Hôpital's Rule, which replaces the given quotient with a quotient of derivatives.

L'Hôpital's Rule (For Forms of Type $\frac{0}{0}$)

Let $\lim\limits_{x \to a} f(x) = \lim\limits_{x \to a} g(x) = 0$

If $\lim\limits_{x \to a} \frac{f'(x)}{g'(x)}$ exists (the limit may be a finite number or $+\infty$ or $-\infty$),

then $\lim\limits_{x \to a} \frac{f(x)}{g(x)} = \lim\limits_{x \to a} \frac{f'(x)}{g'(x)}$ (a may stand for a^-, a^+, $-\infty$, or $+\infty$)

For example, $\lim\limits_{x \to 0} \frac{\sin x}{x} = \lim\limits_{x \to 0} \frac{(\sin x)'}{(x)'} = \lim\limits_{x \to 0} \cos x = \cos 0 = 1$

$\lim\limits_{x \to 0} \frac{1-\cos x}{x} = \lim\limits_{x \to 0} \frac{(1-\cos x)'}{(x)'} = \lim\limits_{x \to 0} -(-\sin x) = \sin 0 = 0$

$\lim\limits_{x \to 2} \frac{x^2-5x+6}{x^2-4} = \lim\limits_{x \to 2} \frac{(x^2-5x+6)'}{(x^2-4)'} = \lim\limits_{x \to 2} \frac{2x-5}{2x} = \frac{2 \cdot 2-5}{2 \cdot 2} = -\frac{1}{4}$

Example

(1) Compute $\lim\limits_{x \to 0} \frac{\tan 3x}{\ln(1+x)}$

Both numerator and denominator have limit 0.

By L'Hôpital's Rule, $\lim\limits_{x \to 0} \frac{\tan 3x}{\ln(1+x)} = \lim\limits_{x \to 0} \frac{(\tan 3x)'}{\{\ln(1+x)\}'} = \lim\limits_{x \to 0} \frac{3 \sec^2 3x}{\frac{1}{1+x}} = \frac{3 \cdot 1}{\frac{1}{1+0}} = 3$

(2) Compute $\lim\limits_{x \to 0} \frac{\sin x - x}{x^3}$

By L'Hôpital's Rule,

$\lim\limits_{x \to 0} \frac{\sin x - x}{x^3} = \lim\limits_{x \to 0} \frac{(\sin x - x)'}{(x^3)'} = \lim\limits_{x \to 0} \frac{\cos x - 1}{3x^2} = \lim\limits_{x \to 0} \frac{(\cos x - 1)'}{(3x^2)'} = \lim\limits_{x \to 0} \frac{-\sin x}{6x}$

$= \lim\limits_{x \to 0} \frac{(-\sin x)'}{(6x)'} = \lim\limits_{x \to 0} \frac{-\cos x}{6} = -\frac{1}{6}$

2. L'Hôpital's Rule for Forms of Type $\frac{\infty}{\infty}$

Consider the limit: $\lim\limits_{x \to \infty} \frac{e^{-x}}{x^{-1}}$

Since the limit has the form $\frac{0}{0}$, we may apply L'Hôpital's Rule indefinitely.

$$\lim_{x\to\infty} \frac{e^{-x}}{x^{-1}} = \lim_{x\to\infty} \frac{(e^{-x})'}{(x^{-1})'} = \lim_{x\to\infty} \frac{-e^{-x}}{-x^{-2}} = \lim_{x\to\infty} \frac{e^{-x}}{x^{-2}} = \lim_{x\to\infty} \frac{(e^{-x})'}{(x^{-2})'} = \lim_{x\to\infty} \frac{-e^{-x}}{-2x^{-3}} = \lim_{x\to\infty} \frac{e^{-x}}{2x^{-3}}$$

$$= \lim_{x\to\infty} \frac{(e^{-x})'}{(2x^{-3})'} = \cdots\cdots\cdots$$

In this case, we have to do algebra first. That is, $\lim_{x\to\infty} \dfrac{e^{-x}}{x^{-1}} = \lim_{x\to\infty} \dfrac{\frac{1}{e^x}}{\frac{1}{x}} = \lim_{x\to\infty} \dfrac{x}{e^x}$

Note that $e^0 = 1$

$\therefore \quad \dfrac{e^x - e^0}{x - 0} = \dfrac{e^x - 1}{x}$

$\therefore \quad \lim_{x\to 0} \dfrac{e^x - 1}{x} = \lim_{x\to 0} \dfrac{(e^x - 1)'}{x'} = \lim_{x\to 0} e^x = e^0 = 1$

L'Hôpital's Rule (For Forms of Type $\dfrac{\infty}{\infty}$)

Let $\lim_{x\to a} |f(x)| = \lim_{x\to a} |g(x)| = \infty$

If $\lim_{x\to a} \dfrac{f'(x)}{g'(x)}$ exists (the limit may be a finite number or $+\infty$ or $-\infty$),

then $\lim_{x\to a} \dfrac{f(x)}{g(x)} = \lim_{x\to a} \dfrac{f'(x)}{g'(x)}$ (a may stand for a^-, a^+, $-\infty$, or $+\infty$)

For example,

Since $\lim_{x\to\infty} e^x = \infty$ and $\lim_{x\to\infty} x = \infty$, by L'Hôpital's Rule, $\lim_{x\to\infty} \dfrac{x}{e^x} = \lim_{x\to\infty} \dfrac{(x)'}{(e^x)'} = \lim_{x\to\infty} \dfrac{1}{e^x} = 0$

Example

(1) Compute $\lim_{x\to\infty} \dfrac{x^a}{e^x}$ for any positive real number a.

Both numerator and denominator have limit ∞.

By L'Hôpital's Rule,

$$\lim_{x\to\infty} \frac{x^a}{e^x} = \lim_{x\to\infty} \frac{(x^a)'}{(e^x)'} = \lim_{x\to\infty} \frac{ax^{a-1}}{e^x} = \lim_{x\to\infty} \frac{(ax^{a-1})'}{(e^x)'} = \lim_{x\to\infty} \frac{a(a-1)x^{a-2}}{e^x}$$

$$= \lim_{x\to\infty} \frac{(a(a-1)x^{a-2})'}{(e^x)'} = \lim_{x\to\infty} \frac{a(a-1)(a-2)x^{a-3}}{e^x} = \lim_{x\to\infty} \frac{a(a-1)(a-2)}{x^{3-a}e^x} = 0$$

(2) Compute $\lim_{x\to\infty} \dfrac{\ln x}{x^a}$ for any positive real number a.

Both numerator and denominator have limit ∞.

By L'Hôpital's Rule,

$$\lim_{x\to\infty} \frac{\ln x}{x^a} = \lim_{x\to\infty} \frac{(\ln x)'}{(x^a)'} = \lim_{x\to\infty} \frac{\frac{1}{x}}{ax^{a-1}} = \lim_{x\to\infty} \frac{1}{ax^a} = 0$$

(3) Compute $\displaystyle\lim_{x\to 0^+} \frac{\ln x}{\cot x}$

Since $\displaystyle\lim_{x\to 0^+} \ln x = -\infty$ and $\displaystyle\lim_{x\to 0^+} \cot x = \infty$, by L'Hôpital's Rule,

$$\lim_{x\to 0^+} \frac{\ln x}{\cot x} = \lim_{x\to 0^+} \frac{(\ln x)'}{(\cot x)'} = \lim_{x\to 0^+} \frac{\frac{1}{x}}{-\csc^2 x} = \lim_{x\to 0^+} \frac{-\sin^2 x}{x} = \lim_{x\to 0^+} \left(-\sin x \cdot \frac{\sin x}{x}\right) = 0 \cdot 1 = 0$$

> *Note:* If we have forms of type $0 \cdot \infty$ and $\infty - \infty$, then use L'Hôpital's Rule after we have transformed the problem to a $\dfrac{0}{0}$ or $\dfrac{\infty}{\infty}$ form.

Example

(1) $\displaystyle\lim_{x\to\frac{\pi}{2}} \tan x \cdot \ln(\sin x)$ (Form of $0 \cdot \infty$)

Since $\displaystyle\lim_{x\to\frac{\pi}{2}} |\tan x| = \infty$ and $\displaystyle\lim_{x\to\frac{\pi}{2}} \ln(\sin x) = 0$, $\displaystyle\lim_{x\to\frac{\pi}{2}} \tan x \cdot \ln(\sin x)$ is the form of $0 \cdot \infty$

Changing $\tan x$ to $\dfrac{1}{\cot x}$,

$$\lim_{x\to\frac{\pi}{2}} \tan x \cdot \ln(\sin x) = \lim_{x\to\frac{\pi}{2}} \frac{\ln(\sin x)}{\cot x} \left(\text{Form of } \frac{0}{0}\right) = \lim_{x\to\frac{\pi}{2}} \frac{\{\ln(\sin x)\}'}{\{\cot x\}'}$$

$$= \lim_{x\to\frac{\pi}{2}} \frac{\frac{\cos x}{\sin x}}{-\csc^2 x} = \lim_{x\to\frac{\pi}{2}} (-\sin x \cos x) = 0$$

(2) $\displaystyle\lim_{x\to 1^+} \left(\frac{x}{x-1} - \frac{1}{\ln x}\right)$ (Form of $\infty - \infty$)

$$\lim_{x\to 1^+} \left(\frac{x}{x-1} - \frac{1}{\ln x}\right) = \lim_{x\to 1^+} \left\{\frac{x\ln x - (x-1)}{(x-1)\ln x}\right\} \left(\text{Form of } \frac{0}{0}\right) = \lim_{x\to 1^+} \frac{\{x\ln x - (x-1)\}'}{\{(x-1)\ln x\}'}$$

$$= \lim_{x\to 1^+} \frac{1\cdot\ln x + x\cdot\frac{1}{x} - 1}{1\cdot\ln x + (x-1)\cdot\frac{1}{x}} = \lim_{x\to 1^+} \frac{\ln x}{\frac{x\ln x + x - 1}{x}} = \lim_{x\to 1^+} \frac{x\ln x}{x\ln x + x - 1} \left(\text{Form of } \frac{0}{0}\right)$$

$$= \lim_{x\to 1^+} \frac{(x\ln x)'}{(x\ln x + x - 1)'} = \lim_{x\to 1^+} \frac{1\cdot\ln x + x\cdot\frac{1}{x}}{1\cdot\ln x + x\cdot\frac{1}{x} + 1} = \lim_{x\to 1^+} \frac{\ln x + 1}{\ln x + 2} = \frac{0+1}{0+2} = \frac{1}{2}$$

> *Note:* If we have forms of type 0^0, ∞^0, 1^∞, then take logarithm first and then use L'Hôpital's Rule.

Example

(1) $\displaystyle\lim_{x\to 0^+}(x+1)^{\cot x}$ (Form of 1^∞)

Let $y = (x+1)^{\cot x}$

Then, $\ln y = \ln(x+1)^{\cot x} = \cot x \cdot \ln(x+1) = \dfrac{\ln(x+1)}{\tan x}$ $\left(\text{Form of } \dfrac{0}{0}\right)$

$\therefore\ \displaystyle\lim_{x\to 0^+}\ln(x+1)^{\cot x} = \lim_{x\to 0^+}\dfrac{\ln(x+1)}{\tan x} = \lim_{x\to 0^+}\dfrac{\{\ln(x+1)\}'}{\{\tan x\}'} = \lim_{x\to 0^+}\dfrac{\frac{1}{x+1}}{\sec^2 x} = 1$

Since $\ln\left\{\displaystyle\lim_{x\to 0^+}(x+1)^{\cot x}\right\} = \lim_{x\to 0^+}\{\ln(x+1)^{\cot x}\} = 1,\quad \lim_{x\to 0^+}(x+1)^{\cot x} = e^1$

Therefore, $\displaystyle\lim_{x\to 0^+}(x+1)^{\cot x} = e$

(2) $\displaystyle\lim_{x\to\frac{\pi}{2}^-}(\tan x)^{\cos x}$ (Form of ∞^0)

Let $y = (\tan x)^{\cos x}$

Then, $\ln y = \ln(\tan x)^{\cos x} = \cos x \cdot \ln(\tan x) = \dfrac{\ln(\tan x)}{\sec x}$

$\therefore\ \displaystyle\lim_{x\to\frac{\pi}{2}^-}\ln(\tan x)^{\cos x} = \lim_{x\to\frac{\pi}{2}^-}\dfrac{\ln(\tan x)}{\sec x} = \lim_{x\to\frac{\pi}{2}^-}\dfrac{\{\ln(\tan x)\}'}{(\sec x)'} = \lim_{x\to\frac{\pi}{2}^-}\dfrac{\frac{\sec^2 x}{\tan x}}{\sec x \tan x}$

$\qquad = \displaystyle\lim_{x\to\frac{\pi}{2}^-}\dfrac{\sec^2 x}{\sec x \tan^2 x} = \lim_{x\to\frac{\pi}{2}^-}\dfrac{\sec x}{\tan^2 x} = \lim_{x\to\frac{\pi}{2}^-}\dfrac{\cos x}{\sin x} = 0$

Since $\ln\left\{\displaystyle\lim_{x\to\frac{\pi}{2}^-}(\tan x)^{\cos x}\right\} = \lim_{x\to\frac{\pi}{2}^-}\{\ln(\tan x)^{\cos x}\} = 0,\quad \lim_{x\to\frac{\pi}{2}^-}(\tan x)^{\cos x} = e^0 = 1$

4-5 Applications of Differentiation

1. Velocity and Acceleration

Velocity has a sign associated with it.
Speed is the absolute value of the velocity.

(1) Velocity

At each time t, the position of the object is described by a number $f(t)$.

Velocity tells how fast the position is changing, and in which direction.

Average velocity from time t_1 to t_2 is $\dfrac{f(t_2)-f(t_1)}{t_2-t_1}$.

Instantaneous velocity $V(t)$ is the limit of the average velocity.

Thus, $V(t) = \displaystyle\lim_{t_2\to t_1}\dfrac{f(t_2)-f(t_1)}{t_2-t_1}$

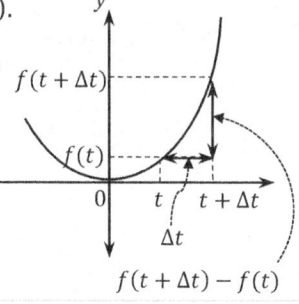

Therefore, the velocity $V(t)$ at time t is the derivative of the position function $f(t)$; i.e.,

$$V(t) = f'(t) = \lim_{\Delta t \to 0} \frac{f(t+\Delta t) - f(t)}{(t+\Delta t) - t}$$

(2) Acceleration

$\circ \circ \circ$ $a(t) = V'(t) = f''(t)$

The difference $f(t + \Delta t) - f(t)$ is the change of position.

$V(t) = f'(t)$ is the rate of change.

Acceleration $A(t)$ is defined to be the rate of change of velocity with respect to time.

That is, $A(t) = \dfrac{d}{dt} V(t) = \dfrac{d^2}{dt^2} f(t)$

Example From the top of a building 50 feet high, an object was thrown upward with an initial velocity of 35 feet per second. The position of the object after t seconds is given by $f(t) = 40 + 35t - 5t^2$.

(1) What was the velocity of the object at $t = 2$, $t = 5$?

(2) With what velocity did the object hit the ground?

(3) When did the object reach its maximum height?

(4) What was its maximum height?

(5) What was the average velocity until the object hit the ground?

(6) What was its acceleration at $t = 3$?

(1) Let $t = 0$ correspond to the instant when the object was thrown, and $V(t)$ be the velocity at time t. Then, $V(t) = \dfrac{d}{dt} f(t) = 35 - 10t$

∴ The velocity at $t = 2$ is $35 - 10 \cdot 2 = 15$ (feet/seconds) and

the velocity at $t = 5$ is $35 - 10 \cdot 5 = -15$ (feet/seconds)

(2) The object hit the ground when $f(t) = 0$.

∴ $40 + 35t - 5t^2 = 0$; $t^2 - 7t - 8 = 0$; $(t + 1)(t - 8) = 0$

∴ $t = -1$ or $t = 8$

Since $t > 0$, $t = 8$

∴ $V(8) = 35 - 10 \cdot 8 = -45$ (feet/seconds)

(3) The object reached its maximum height at the time its velocity was 0.

∴ $V(t) = 35 - 10t = 0$; $t = \dfrac{7}{2} = 3.5$

After 3.5 seconds, the object reaches its maximum height.

(4) The maximum height is $f(3.5) = 40 + 35(3.5) - 5(3.5)^2 = 101.25$ feet

(5) Since the total taken time is 8 seconds and

total distance of the movement is $101.25 \times 2 - 50 = 152.5$ feet,

the average velocity is $\dfrac{152.5}{8} \approx 19.06$ (feet/seconds)

(6) $A(t) = V'(t) = -10$

2. Differentiation of Vector Functions

(1) Velocity Vector

Consider a point moving in the plane.

At each time t, the point is at some point (x, y).

Let $x = f(t)$, $y = g(t)$.

It is convenient to think of the ordered pair $\langle f, g \rangle$ as a vector function **u**.

$\mathbf{u}(t) = \langle f(t), g(t) \rangle$ is called the *position vector*.

If a point moves counterclockwise around the circle $x^2 + y^2 = r^2$ at a constant speed of one

revolution per second, starting at $(r, 0)$ when $t = 0$, then $\mathbf{u}(t) = \langle r \cos 2\pi t, \ r \sin 2\pi t \rangle$.

The motion of an object at a time is described by its speed and direction (i.e., by a vector V),

called the *velocity vector*.

The direction of V is the direction of motion and the magnitude of V is the speed.

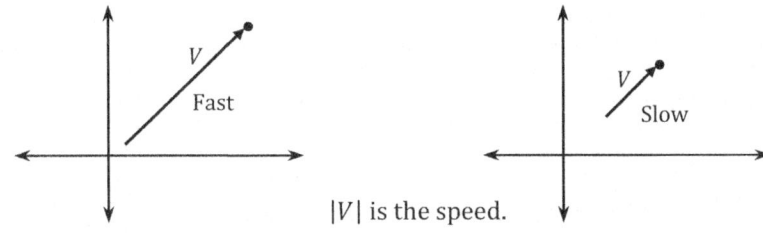

$|V|$ is the speed.

The change in position from time t to time $t + h$ is $\mathbf{u}(t + h) - \mathbf{u}(t)$.

The *average velocity* over the interval $[t, t + h]$ is $\boxed{V_{ave} = \dfrac{\mathbf{u}(t+h) - \mathbf{u}(t)}{h} = \dfrac{\text{change in position}}{\text{change in time}}.}$

Taking the limit as $h \to 0$, we have the *instantaneous velocity* at time t; i.e.,

$$\boxed{V = V(t) = \lim_{h \to 0} \frac{\mathbf{u}(t+h) - \mathbf{u}(t)}{h} = \mathbf{u}'(t) \quad \text{(Derivative of the position function } \mathbf{u})}$$

The *instantaneous speed* is the magnitude of the velocity vector.

That is, instantaneous speed is $A(t) = |V(t)| = |\mathbf{u}'(t)|$.

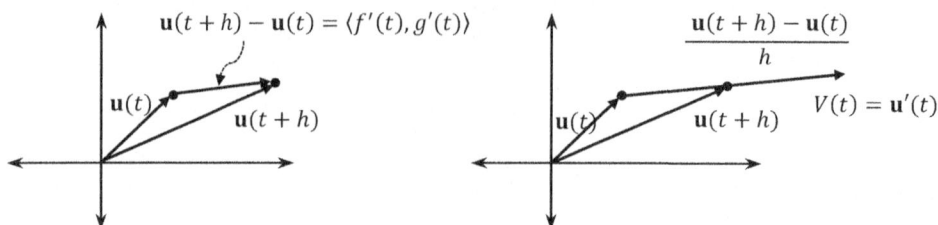

For example,

1) Constant velocity implies straight line motion.

$$\mathbf{u}(t) = \langle x + at, \ y + bt \rangle \ \Rightarrow \ \mathbf{u}'(t) = \langle a, \ b \rangle \qquad \therefore V(t) = \langle a, \ b \rangle$$

2) $\mathbf{u}(t) = \langle r\cos t, \ r\sin t \rangle \ \Rightarrow \ V(t) = \mathbf{u}'(t) = \langle -r\sin t, \ r\cos t \rangle$

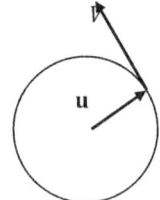

The speed of the motion is the length of the velocity vector.

That is, $A(t) = |V(t)| = \sqrt{r^2 \sin^2 t + r^2 \cos^2 t} = r$ (Constant)

(2) Acceleration Vector

The velocity of a point P moving in the plane
is a vector whose components are the rates of change of
the x- and y-coordinates of P with respect to time.

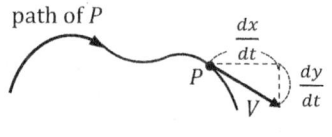

path of P

Consider the Newton's law for motion in the plane.

Let $F(t)$ be the total force acting on an object at time t.

The force acts with a certain strength in a certain direction. Thus, F is a vector function.

Newton's law

$F = (mV)'$ where m is mess of an object and V is its velocity.

Since the mass is independent of t, it may be considered as a constant, so $F = mV'$

The derivative V' is called the acceleration, denoted by A.

Thus, $A = V' = \mathbf{u}''$

Note:
$$x = f(t), \ y = g(t)$$
$$V(t) = \langle \frac{dx}{dt}, \ \frac{dy}{dt} \rangle = \langle f'(t), \ g'(t) \rangle$$
$$|V(t)| = \sqrt{\left(\frac{dx}{dt}\right)^2 + \left(\frac{dy}{dt}\right)^2} = \sqrt{\{f'(t)\}^2 + \{g'(t)\}^2}$$

$$|A(t)| = \sqrt{\frac{d^2x}{dt^2} + \frac{d^2y}{dt^2}} = \langle f''(t),\ g''(t)\rangle$$

Example   Parametric equations for a point P moving in the plane are

$$x = 2t + 1,\ y = 2t - t^2,\ \text{where } t \text{ represents time.}$$

(1) Graph the path of P.

$$x = 2t + 1 \ \Rightarrow\ t = \frac{x-1}{2}$$

$$\therefore\ y = 2t - t^2 = 2\left(\frac{x-1}{2}\right) - \left(\frac{x-1}{2}\right)^2 = x - 1 - \frac{(x-1)^2}{4}$$

$$= \frac{-x^2 + 6x - 5}{4} = -\frac{1}{4}(x-3)^2 + 1$$

Since $t \geq 0$, $\frac{x-1}{2} \geq 0$ $\therefore\ x \geq 1$

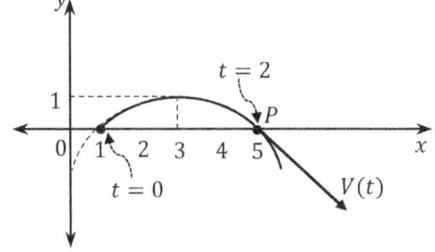

The path is the part of a parabola shown in the Figure.

(2) Find the expressions for the velocity $V(t)$, speed $|V(t)|$, and acceleration $A(t)$.

$$\frac{dx}{dt} = 2;\ \frac{dy}{dt} = 2 - 2t \qquad \therefore\ V(t) = \langle 2, 2 - 2t\rangle$$

$$|V(t)| = \sqrt{\left(\frac{dx}{dt}\right)^2 + \left(\frac{dy}{dt}\right)^2} = \sqrt{2^2 + (2-2t)^2} = \sqrt{4t^2 - 8t + 8} = 2\sqrt{t^2 - 2t + 2}$$

$$\frac{d^2x}{dt^2} = 0;\ \frac{d^2y}{dt^2} = -2 \qquad \therefore\ A(t) = \langle 0, -2\rangle$$

(3) Find the direction of the velocity vector at $t = 2$.

$$V(t) = \langle 2, 2 - 2t\rangle$$

$\therefore$ At $t = 2$, $V(2) = \langle 2, -2\rangle$

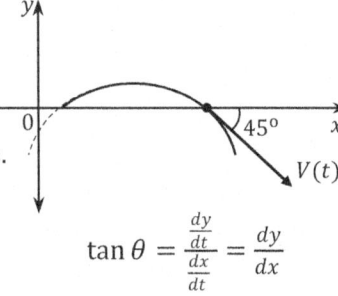

Let θ be the angle between the velocity vector $V(t)$ and x-axis.

Then, $\tan\theta = -1$ $\therefore\ \theta = -45°$

$$\tan\theta = \frac{\frac{dy}{dt}}{\frac{dx}{dt}} = \frac{dy}{dx}$$

Exercises

#1 Find the values of constants $a, b,$ and c.

(1) For the graph of a function $y = x^3 + ax + b$, the equation of the tangent line at $(1, 3)$ on the graph is $y = 2x + c$.

(2) For the graph of a function $y = ax^3 + bx^2 + cx$ and two points $(1, 0)$ and $(2, 4)$ on the graph, the tangent lines at the points are parallel to each other.

(3) For the graph of $y = x^3 + ax^2 + bx + c$ passing through $(1, 2)$, the graph and the line $y = 3x - 2$ intersect only at a point $(2, 4)$.

(4) The two graphs of $y = a + \sin x$ and $y = \cos^2 x + bx$ intersect only at one point on a line $y = x + c$. $\left(-\dfrac{\pi}{2} < x < \dfrac{\pi}{2} \right)$

#2 Find the equation of the tangent line to the graph at the point indicated.

(1) $y = x^2$ at $(2, 4)$

(2) $y = x^3 + x^2 - 3x + 4$ at $(1, 5)$

(3) $y = \dfrac{\ln x}{x}$ at $(0, 0)$

(4) $x^2 + y^2 = r^2$ at (x_1, y_1)

(5) $y^2 = 4px$ at (x_1, y_1)

(6) $\dfrac{x^2}{a^2} + \dfrac{y^2}{b^2} = 1$ at (x_1, y_1)

(a, b ; constants)

(7) $x \cos y + y \cos x = -2\pi$ at (π, π)

(8) $x^2 + 5ye^x + y^3 = 4$ at $(0, 1)$

(9) $y = x^{2x}$ $(x > 0)$ at $(1, 1)$

(10) $x = \cos^3 \theta,\ y = \sin^3 \theta$ at $\theta = \dfrac{\pi}{6}$

(11) $y = |x^2 - x[x]|$ at $x = 2.5$

($[x]$; The greatest integer in x)

#3 Find the equation of the normal line to the curve at the point indicated.

(1) $y = x^3$ at $(-1, -1)$

(2) $y = x^3 - 3x^2 + 4$ at $(1, 2)$

(3) $y = e^{x-1}$ at $(3, e)$

(4) $y = \log(1 + x)$ at $(1, 0)$

(5) $y = \dfrac{x}{1-x}$ at $(3, 1)$

(6) $y = \tan^2 x$ at $\left(\dfrac{\pi}{4}, 1 \right)$

#4 Find the value.

(1) When the graphs of $y = ax^2$ and $y = \log x$ intersect only at one point, find the value of the constant a.

(2) When the two graphs of $y = a - 2\cos^2 x$ and $y = 2\sin x$ intersect only at one point, find the value of the constant a. $(0 < x < 2\pi)$

(3) Find the value of the positive constant a so that the tangent lines of the two graphs $f(x) = 3x^2$ and $g(x) = a - 4x^2$ are perpendicular to each other.

(4) When the tangent lines of the two graphs $y = \log(2x + 3)$ and $y = a - \log x$ are perpendicular to each other, find the value of the constant a.

(5) Find the equation of the tangent line at $\left(\frac{1}{2}, \frac{1}{2}\right)$ to the curve $x^3 + y^3 - 3xy = 0$.

(6) When the tangent line at $(1, e)$ on the graph of $y = e^x$ intersects only at one point on the graph of $y = 3\sqrt{x - a}$, find the value of the real number a.

(7) For the tangent line at $(0, 0)$ on the graph of $y = \frac{2}{3}\sin 2x + 3x^2$, let θ be the angle between the tangent line and x-axis. Find the slope of the line which bisects the angle. $(0 < \theta < \frac{\pi}{2})$

(8) Let $P(0, y_1)$ and $Q(3, y_2)$ be points on the graph of $y = x^3 - 4x$. For the line l passing through the two points P and Q, find the equation of the tangent line, which is parallel to the line l, of the graph.

(9) When the graph of $y = x^3 - 4x + 3$ has two tangent lines which are parallel to the line $y = 2x + 3$, find the distance between the two tangent lines.

(10) When the graph of $y = x^4$ and a circle with center on y-axis intersect only at one point $(1, 1)$, find the value of the radius of the circle.

(11) When the graph of $y = x^2 + 3$ has two tangent lines at a point $P(1, 0)$, and Q and R are the points of tangency, find the area of the triangle $\triangle PQR$.

(12) When the tangent line at $P(a, e^{2a})$ on the graph of $y = e^{2x}$ crosses the x-axis at Q, let R be the point on x-axis at which the line segment $\overline{PR}$ and x-axis are perpendicular. Find the value of a so that the area of the triangle $\triangle PQR$ is 64.

#5 For a function $f(x) = e^{2x-3}$, let $g(x)$ be the inverse of $f(x)$.
Find the equation of the tangent line at $x = 1$ on the graph of $y = g(x)$.

#6 Find the number c guaranteed by the Mean Value Theorem for the following function on the given interval.

(1) $f(x) = x^2$, $[0, 2]$

(2) $f(x) = 2\sqrt{x - 1}$, $[1, 3]$

(3) $f(x) = \log x$, $[1, e]$

#7 Find the number c guaranteed by the Rolle's Theorem for the continuous function $f(x) = e^x + e^{-x}$ on a closed interval $[-2, 2]$.

#8 Find the limit using the Mean Value Theorem.

(1) $\lim\limits_{x \to 2} \dfrac{3^x - 3^2}{x - 2}$

(2) $\lim\limits_{x \to 0^+} \dfrac{\sin x - \sin(\sin x)}{x - \sin x}$

#9 For $x > 0$, prove the inequality.

(1) $\dfrac{1}{x+1} < \log(x + 1) - \log x < \dfrac{1}{x}$

(2) $0 < \dfrac{1}{x} \log \dfrac{e^x - 1}{x} < 1$

#10 Use L'Hôpital's Rule to complete the following limits.

(1) $\lim\limits_{x \to a} \dfrac{a \sin x - x \sin a}{x - a}$

(2) $\lim\limits_{x \to 0^+} \left(\dfrac{1}{x} - \dfrac{1}{\sin x} \right)$

(3) $\lim\limits_{x \to 0} \dfrac{1 - \cos x - \frac{1}{2} x^2}{x^4}$

(4) $\lim\limits_{x \to 0} \dfrac{x - \log(1+x)}{x^2}$

(5) $\lim\limits_{x \to \infty} \dfrac{e^{-x}}{x^{-2}}$

(6) $\lim\limits_{x \to \infty} \dfrac{e^{3x}}{x^3}$

(7) $\lim\limits_{x \to 0} \dfrac{a^x - b^x}{x}$ $(a > 0, \ b > 0)$

(8) $\lim\limits_{x \to \infty} \left(1 + \dfrac{1}{x} \right)^x$

(9) $\lim\limits_{x \to 0} (\cos x)^{\frac{1}{x^2}}$

#11 Show that each of the following is increasing on $(-\infty, \infty)$.

(1) $f(x) = x^3 - 6x^2 + 15x + 10$

(2) $f(x) = x^3 - 6x^2 + 12x - 8$

(3) $f(x) = x - \sin x$

#12 Prove the statement.

(1) A function $y = e^x - x$ is increasing on $[0, \infty)$.

(2) A function $y = x^{\frac{1}{x}}$ is decreasing on $[3, \infty)$.

#13 Find the value of the real number a so that the following statement is true.

(1) A function $f(x) = x^3 + ax^2 + ax + 10$ is increasing on $(-\infty, \infty)$.

(2) A function $f(x) = x^3 - 6x^2 + ax - 2$ is decreasing on $(0, 3)$.

(3) A function $f(x) = 2x^3 - 3ax^2 + 6ax - 6x - 1$ is decreasing on $[1, 3]$.

(4) A function $f(x) = ax + \sin x$ is increasing on $(-\infty, \infty)$.

(5) A function $f(x) = (ax^2 + 1)e^x$ is increasing on $(-\infty, \infty)$.

(6) A function $f(x) = (a - x)e^{x^2}$ is decreasing on $(-\infty, \infty)$.

#14 Find all local extreme values of the graph of $f(x)$.

(1) $f(x) = x^3 - 3x - 2$

(2) $f(x) = -x^3 + 12x - 3$

(3) $f(x) = x^3 - 3x^2 + 3x + 1$

(4) $f(x) = 2x^3 + 3x^2 - 36x - 40$

(5) $f(x) = -2x^3 + 9x^2 - 12x + 5$

(6) $f(x) = \frac{1}{2}x^4 - \frac{4}{3}x^3 - x^2 + 4x + 5$

(7) $f(x) = |x - 1|$

(8) $f(x) = \sqrt[3]{x^2}$

(9) $f(x) = \frac{2x-1}{x^2+6}$

(10) $f(x) = \frac{2-\cos x}{\sin x}$ $(0 < x < \frac{\pi}{2})$

(11) $f(x) = \frac{\log x}{x^4}$

#15 State if the following statement is true or false.

(1) For a function $f(x) = xe^x$,

   1) $f(x)$ has a local minimum value at $x = 1$.

   2) $f(x)$ is decreasing on $x < -1$

   3) $f(x)$ is concave upward on the interval $(-\infty, -2)$.

(2) For a function $f(x) = x^{\sqrt{x}}$ $(x > 0)$,

   1) $f(x)$ is increasing on $x \geq 1$.

   2) $f(x) \geq 1$ for all $x \geq 1$.

   3) The equation of the function $f(x) = 2$ has two different real number solutions when $x \geq 1$.

   4) $f'(x) \geq 0$ for all $x > 0$.

(3) For a function defined on an open interval $(0, 12)$, the derivative $y = f'(x)$ is shown as the Figure.

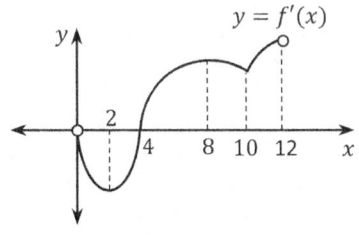

   1) $f(x)$ is increasing on $(4, 12)$.

   2) $f(x)$ has a local maximum value at $x = 4$.

   3) $f(x)$ has minimum value at $x = 2$ in the open interval $(0, 4)$

   4) $f(x)$ is not differentiable at $x = 10$.

(4) When a function $f(x)$ with second derivative satisfies $f(-x) = -f(x)$ for all real number x,

   1) $f'(-x) = -f'(x)$

   2) $\lim_{x \to 0} f'(x) = -1$

3) If $f'(x)$ has a local maximum value at $x = a$ ($a \neq 0$), then $f'(x)$ has a local minimum value at $x = -a$.

(5) For a function $f(x) = x + \sin x$, let $g(x) = (f \circ f)(x)$.

   1) The graph of $f(x)$ is concave upward on an open interval $(0, \pi)$.

   2) $g(x)$ is increasing on $(0, \pi)$.

   3) There is x $(0 < x < \pi)$ such that $g'(x) = 0$.

(6) For any polynomial function $f(x)$ with a local maximum value at $x = 0$,

   1) $|f(x)|$ has a local maximum value at $x = 0$.

   2) $f(|x|)$ has a local maximum value at $x = 0$.

   3) $f(x) - x^2|x|$ has a local maximum value at $x = 0$.

#16 Find the local extreme values using the second derivatives.

(1) $f(x) = x^2 e^{-x}$

(2) $f(x) = \sqrt{3}\sin x + \cos x$ $(0 \leq x \leq 2\pi)$

(3) $f(x) = 2x - \tan x$ $(0 < x < \pi)$

#17 Find the value.

(1) For a function $f(x) = x^3 + ax^2 + ax + 1$,

   1) Find the range of a so that $f(x)$ has local maximum and minimum values.

   2) Find the range of a so that $f(x)$ does not have local extreme values.

(2) When a function $f(x) = x^3 - 3x + a$ has a local maximum value 12,
find the value of a and local minimum value.

(3) When a function $f(x) = x^3 + ax^2 + bx + 3$ has a local minimum value 2 at $x = 1$,
find the values of a and b.

(4) A function $f(x) = x^3 + 3ax^2 + bx + c$ has a local maximum value at $x = 1$ and a local
minimum value at $x = 3$.

   1) Find the values of a and b.

   2) Find the local extreme values.

   3) When the local maximum value is 10, find the value of c and the local minimum value.

(5) Find the range of a so that a function $f(x) = x^4 - 4x^3 + 6ax^2$ has no local extreme values.

(6) When a function $f(x) = x^3 + \left(a - \dfrac{1}{2}\right)x^2 - 2(a+1)x + a$ has a local maximum value at $x = 1$, find the range of the real number a.

(7) When a function $f(x) = 2x^3 + 3ax^2$ has a local minimum value in the open interval $(1, 2)$, find the range of the real number a.

(8) When a differentiable function $f(x)$ defined on real numbers satisfies:

    i) $f(x - y) = f(x) - f(y) + xy(x - y)$ for any real numbers, and

    ii) $f'(0) = 10$,

    $f(x)$ has a local maximum value at $x = a$ and local minimum value at $x = b$.

    Find the values of a and b.

(9) When a cubic function $y = f(x)$ with integer coefficients satisfies:

    i) $f(-x) = -f(x)$ for all x,

    ii) $f(1) = 5$, and

    iii) $1 < f'(1) < 6$,

    Find the sum of the local extreme values.

(10)   When a function $f(x) = \log x + \dfrac{a}{x} - x$ has local extreme values, find the range of a.

(11)   For a function $f(x) = x^3 - 2x^2 \sin\theta + x + 1 \quad \left(0 \le \theta \le \dfrac{\pi}{2}\right)$,

    1)   Find the range of θ at which $f(x)$ has local extreme values.

    2)   Find the range of θ at which $f(x)$ has no local extreme values.

(12)   When a function $f(x) = ax + 2\sin x$ has local extreme values, find the range of a.

(13)   When a function $f(x) = e^{ax} \sin x$ has a local maximum value at $x = \dfrac{\pi}{4}$,

    find the value of a and local maximum value.

#18   Find the cubic function such that:

    i)   a point $(0, 2)$ is the inflection point of the curve $y = f(x)$,

    ii)   the tangent line at $(0, 2)$ and a line $y = -2x$ are parallel each other, and

    iii) $f(x)$ has a local minimum value at $x = 1$.

#19   Find the inflection points for the following function.

    (1) $f(x) = x^4 - 4x^3 + 4$ (2) $y = \left(\ln\dfrac{1}{ax}\right)^2$

#20   Find the equation of the tangent line at the inflection point of the function.

    (1) $y = \sin^2 x \quad \left(0 \le x \le \dfrac{\pi}{2}\right)$ (2) $y = xe^{-x}$

#21 When the graph of $y = \log(ax^2 + b)$ has an inflection point $(1, \log 4)$,

find the values of a and b.

#22 Find the maximum value of the integer a at which a function

$f(x) = -x^3 + 3(a-1)x^2 + 3(a-3)x + 2$ has no local extreme values. $(x \geq 0)$

#23 Find the maximum and minimum values for each function.

(1) $f(x) = x^3 - \frac{9}{2}x^2 + 6x - \frac{1}{2}$ $(0 \leq x \leq 3)$

(2) $f(x) = |x^2(x-3)|$ $(-1 \leq x \leq 5)$

(3) $f(x) = x + \sqrt{1 - 2x - x^2}$

(4) $f(x) = x - 3 + \sqrt{9 - x^2}$

(5) $f(x) = \frac{x}{x^2 - x + 1}$

(6) $f(x) = 2\sin^3 x + 3\cos^2 x$ $(0 \leq x \leq 2\pi)$

(7) $f(x) = \sin x (1 + \cos x)$ $(0 \leq x \leq \pi)$

(8) $f(x) = \sin x + \sin 2x \cos x + \frac{3}{4}\cos 2x$ $(0 \leq x \leq 2\pi)$

(9) $f(x) = \sin x + \sqrt{3}\cos x + x$ $(0 \leq x \leq \pi)$

(10) $f(x) = \sin x + |\cos x|$ $(0 \leq x \leq 2\pi)$

(11) $f(x) = (\sin x + \cos x)^3 - 12\sin x \cos x$

(12) $f(x) = \frac{\sin x + \cos x}{\sin x + \cos x - 2}$

(13) $f(x) = (x^2 - 8)e^x$ $(-3 \leq x \leq 3)$

(14) $f(x) = \sqrt{6 - x^2}e^x$

#24 Find the value.

(1) When a function $f(x) = ax^4 - 4ax^3 + b$ $(a > 0)$ on a closed interval $[1, 4]$ has the

maximum and minimum values 3 and -15, respectively, find the values of a and b.

(2) When a function $f(x) = \frac{x+1}{x^2+8}$ $(0 \leq x \leq a)$ has maximum and minimum values $\frac{1}{4}$ and

$\frac{1}{8}$, respectively, find the maximum value of a.

(3) For a function $y = \sin^2 x$ $\left(0 \leq x \leq \frac{\pi}{2}\right)$, the function is concave downward in the range $a <$

$x \leq b$. Find the values of a and b.

(4) For a function $f(x) = e^{2x}$ on a closed interval $[0, x]$, find the values of a and b so that the

inequality $a < \frac{1}{x} \ln \frac{e^{2x}-1}{2x} < b$ is always true.

(5) When a function $f(x) = x \log x + 3x + a$ has minimum value 1, find the value of a.

#25 Find the range.

(1) For an equation $2x^3 - 6x + a = 0$, find the range of a so that:

    1) The equation has three different real number solutions.

    2) The equation has a double root and a real number solution.

    3) The equation has a real number solution and two complex number solutions.

(2) When an equation $x^3 - 3x^2 - 24x - a = 0$ has two different positive number solutions and

a negative number solution, find the range of a.

(3) Find the range of a so that $-x^3 + 3x^2 - a = 0$ has a solution which is greater than 5.

(4) When an equation $3x^4 - 4x^3 - 12x^2 - a = 0$ has exactly two different negative real

number solutions, find the range of a.

(5) When an equation $x^4 - 4a^3x + 27 = 0$ has no real number solution, find the range of a.

(6) When a function $f(x) = x^3 - 3ax - 2a$ has local extreme values and an equation $f(x) = 0$

has only one real number solution, find the range of a.

(7) Find the range of a so that a function $y = \frac{1}{2}x^4 - 3x^2 + 2ax$ has local extreme values.

(8) When the curve $y = e^{ax}$ and a line $y = x$ intersect at two different points,

    find the range of a.

(9) For a curve $y = 2 \sin x$ $(0 \leq x \leq \pi)$ and a line $y = x + a$, find the range of a so that there

is at least one intersection point.

(10) When a curve $y = x^3 - 3x$ has exactly one tangent line at $(-1, a)$, find the range of a.

(11) For a curve $y = x^3 - x + 1$ and a line $y = 11x + a$, find the range of a so that:

    1) The curve and the line intersect at three different points.

    2) The curve and the line intersect at two points. At one of the points, the line is the

       tangent of the curve.

    3) The curve and the line intersect only at a point.

(12) When an inequality $x^4 - 4x - a^2 + 4a > 0$ is always true for all real number x,

    find the range of a.

(13) For any real number x such that $x > -1$, the inequality $\frac{4}{3}x^3 - x^2 - 2x + 1 - a > 0$ is

always true. Find the range of a.

(14) For any positive number a, the inequality $\sqrt{x} > a \log x$ is always true. Find the range of a.

(15) For any real number x such that $0 \leq x \leq \frac{\pi}{2}$, $\cos x - a = a \sin x$ is always true. Find the range of a.

#26 Find the value.

(1) When an equation $\log x = x + a$ has two different real number solutions, find the value of a.

(2) When an equation $x \log x - 3x - a = 0$ has exactly one real number solution, find the minimum value of a.

(3) When an inequality $ax \geq \log x$ is always true for all $x > 0$, find the minimum value of a positive number a.

(4) When an inequality $3x^4 + 4x^3 - 12x^2 \geq a - 2\sin\left(\frac{\pi}{2}x\right)$ is always true for any real number x, find the maximum value of a.

#27 An object moves along a horizontal coordinate line. At the end of t seconds its directed distance from the origin, in feet, is given by $S = 48t - 16t^2$. How long does it take to stop?

#28 An object, projected vertically upward with an initial velocity of 100 feet/sec, moves according to the law $S = 100t - 8t^2$, where S is the distance from the starting point.

(1) Find the velocity and acceleration when $t = 6$ and when $t = 7$.

(2) Find the greatest height reached.

(3) When will its height be 200 feet?

#29 An object moves on a horizontal coordinate line. Its directed distance S from the origin at the end of t seconds is $S = -t^3 + at^2 + bt + 1$ feet. When $t = 1$, $S = -3$. When is the object moving to the left?

#30 The positions of two objects P and Q, on a coordinate line at the end of t seconds are given by $S_1 = t^2(t^2 - 6t + 12)$ and $S_2 = mt$, respectively.
Find the value of m so that the objects have the same velocity for three times.

#31 Ships A and B start from the origin at the same time. Ship A travels due east at a rate of 20 miles per hour and ship B travels due north at the rate of 40 miles per hour. Find the speed at the intersection point of the line segment $\overline{AB}$ and a line $y = 3x$.

#32 A ship A is sailing due south at 8 miles/hour and ship B, 16 miles south of A, is sailing due east at 6 miles/hour.

    (1) At what rate are they approaching or separating at the end of 1 hour?

    (2) At the end of 2 hours?

    (3) When do they cease to approach each other and how far apart are they at the time?

#33 An object travels along the x-axis so that its position S is $S = \frac{1}{3}t^3 - 4t^2 + 12t$ meters after t seconds. What is its maximum speed on the interval $2 \le t \le 7$.

#34 An object moves in the plane so that its position is $(e^t \cos t,\ e^t \sin t)$. Find the value of t when the speed of the object is $\sqrt{2}e^2$.

#35 A point P is moving in the plane so that its coordinates after t seconds are $(t - \sin t,\ \cos t)$, measured in feet.

    (1) Find the maximum speed of the point P.

    (2) Find the angle between the velocity vector of P at $t = \frac{\pi}{3}$ and x-axis.

#36 A particle rotates counterclockwise from rest according to the law $\theta = \frac{t^3}{81} - \frac{t}{3}$, where θ is in radians and t in seconds. Find the angular displacement θ, the angular velocity V, and the angular accelation A at the end of 9 seconds.

#37 Suppose that a point P moves around a circle with center $(0, 0)$ and radius r at a constant angular speed of ω radians per second. When its initial position is $(r, 0)$,

    (1) Express x and y at t as t.

    (2) Find expressions for the velocity $V(t)$ and its acceleration $A(t)$.

    (3) Find the speed $|V(t)|$.

    (4) Show that the velocity and acceleration vectors are always perpendicular to each other.

#38 Water is running into a conical reservoir,

6 feet in diameter and 12 feet tall (with vertex down) at a rate of 1 cubic feet per minute.

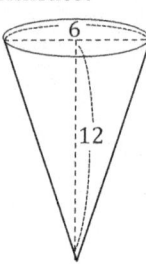

    (1) At what rate is the water level rising

      when the water is 4 feet deep?

    (2) At what rate is the area of the water surface increasing

      when the water is 5 feet deep?

    (3) At what rate is the wetted surface of the reservoir increasing

      when the water is 8 feet deep?

Chapter 5. The Indefinite Integral

5-1 Antiderivatives and Indefinite Integration

1. Antiderivatives

(1) Definition

If $F(x)$ is a function whose derivative $F'(x)$ is $f(x)$, then $F(x)$ is called an *antiderivative* of $f(x)$.

Example   Find an antiderivative of the function $f(x) = 3x^2$.

Consider a function F satisfying $F'(x) = 3x^2$ for all real x.

By differentiation, $F(x) = x^3$.

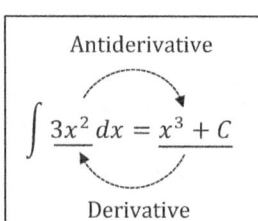

Note that the function $F(x) = x^3 + 2$ also satisfies $F'(x) = 3x^2$.

Thus, $F(x) = x^3 + 2$ is also an antiderivative of $f(x) = 3x^2$.

In fact, $F(x) = x^3 + C$, where C is any constant, is an antiderivative of $3x^2$ on $(-\infty, \infty)$.

That is, every antiderivative of $f(x) = 3x^2$ is of the form $F(x) = x^3 + C$.

(2) Notation for Antiderivatives

Consider the general antiderivative of $f(x) = x^2$ on $(-\infty, \infty)$.

Since its derivative is $F'(x) = \frac{1}{3} \cdot 3x^2 = x^2 = f(x)$, the general antiderivative is

$F(x) = \frac{1}{3}x^3 + C.$

We may write $\frac{d}{dx}(x^2) = D_x(x^2) = \frac{1}{3}x^3 + C.$

But, use the symbol $\int \cdots dx$ rather than $\frac{d}{dx}$ or D_x.

Thus, we write $\int x^2 dx = \frac{1}{3}x^3 + C$ and $\int 3x^2 dx = x^3 + C$.

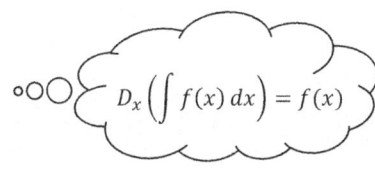

$$\int f(x)dx = F(x) + C, \text{ where } C \text{ is an arbitrary constant.}$$
$$F'(x) = f(x) \iff \int f(x)dx = F(x) + C$$

2. Indefinite Integration

(1) Definition

We use the term indefinite integral in place of antiderivative.

In the symbol $\int f(x)dx$, $\int$ is called the *integral sign* and $f(x)$ is called the *integrand*.

$\int f(x)dx$ is called the *indefinite integral*.

> Antidifferentiate
> $\equiv$ Integrate

Example   Find the function $f(x)$ satisfying $\int f(x)dx = x^3 + 4x^2 + C$.

$f(x) = (x^3 + 4x^2 + C)' = 3x^2 + 8x$

> Indefinite Integral always involves an arbitrary constant.

Note:

$$(1) \int \left(\frac{d}{dx}f(x)\right)dx = f(x) + C$$

$$(2) \frac{d}{dx}\left(\int f(x)dx\right) = f(x)$$

$\because$ (1) Let $\int \left(\frac{d}{dx}f(x)\right)dx = F(x)$

Then, $\frac{d}{dx}F(x) = \frac{d}{dx}f(x)$; $\frac{d}{dx}F(x) - \frac{d}{dx}f(x) = 0$; $\frac{d}{dx}\left(F(x) - f(x)\right) = 0$

$\therefore$ $F(x) - f(x) = C$; $F(x) = f(x) + C$

$\therefore$ $\int \left(\frac{d}{dx}f(x)\right)dx = f(x) + C$

(2) Let $\int f(x)dx = F(x) + C$

Then, $\frac{d}{dx}F(x) = f(x)$

$\therefore$ $\frac{d}{dx}[\int f(x)dx] = \frac{d}{dx}[F(x) + C] = f(x)$

$\therefore$ $\frac{d}{dx}[\int f(x)dx] = f(x)$

Example   Compute the following:

$$(1) \int \left(\frac{d}{dx}x^3\right)dx \qquad (2) \frac{d}{dx}\left(\int x^3dx\right)$$

(1) $\frac{d}{dx}x^3 = 3x^2$

$\therefore$ $\int \left(\frac{d}{dx}x^3\right)dx = \int 3x^2\,dx = x^3 + C$

(2) $(x^4)' = 4x^3$ $\qquad \therefore$ $\frac{1}{4}(x^4)' = x^3$ $\qquad \therefore$ $\int x^3\,dx = \frac{1}{4}x^4 + C$

$\therefore$ $\frac{d}{dx}(\int x^3\,dx) = \frac{d}{dx}\left(\frac{1}{4}x^4 + C\right) = x^3$

(2) Fundamental Integration Formulas

1) For any constant k, $\quad \boxed{\displaystyle\int k\,dx = kx + C}$

$\because$ Since $\dfrac{d}{dx}(kx + C) = k$, $\quad \int k\,dx = kx + C$

The domain of $y = \log x$ is $(0, \infty)$.
$\therefore$ We need the absolute value of x.

2) <u>Power Rule</u>

If n is any rational number except -1, then $\displaystyle\int x^n dx = \dfrac{1}{n+1}x^{n+1} + C$

If $n = -1$, then $\displaystyle\int x^{-1}dx = \int \dfrac{1}{x}dx = \log|x| + C$

$\because$ i) When $n \neq -1$, $\dfrac{d}{dx}\left(\dfrac{1}{n+1}x^{n+1} + C\right) = \dfrac{n+1}{n+1}x^n = x^n$, $\displaystyle\int x^n\,dx = \dfrac{1}{n+1}x^{n+1} + C$

ii) When $n = -1$, $\int \dfrac{1}{x}dx = \int x^{-1}\,dx = \dfrac{1}{-1+1}x^{-1+1} + C = \dfrac{1}{0}x^0 + C$

Since division by 0 is undefined, the power rule for this integration doesn't apply.

Recall that $\dfrac{d}{dx}(\log x) = \dfrac{1}{x}$

Since $\dfrac{d}{dx}(\log|x| + C) = \dfrac{1}{x}$, $\int \dfrac{1}{x}dx = \log|x| + C$

Note: $\boxed{\displaystyle\int (ax + b)^n dx = \dfrac{1}{a}\cdot\dfrac{1}{n+1}(ax+b)^{n+1} + C \quad (a \neq 0,\ n(\geq 0);\text{integer})}$

To antidifferentiate a power of x, we add 1 to the exponent and divide by the new exponent.

3) $\boxed{\displaystyle\int kf(x)dx = k\int f(x)dx}$

$\because \dfrac{d}{dx}(k\int f(x)\,dx) = k\dfrac{d}{dx}(\int f(x)\,dx) = kf(x)$

$\therefore \int kf(x)\,dx = k\int f(x)\,dx$

4) $\boxed{\displaystyle\int \{f(x) + g(x)\}dx = \int f(x)dx + \int g(x)dx}$

$\because \dfrac{d}{dx}(\int f(x)\,dx + \int g(x)\,dx) = \dfrac{d}{dx}(\int f(x)\,dx) + \dfrac{d}{dx}(\int g(x)\,dx) = f(x) + g(x)$

$\therefore \int\{f(x) + g(x)\}\,dx = \int f(x)\,dx + \int g(x)\,dx$

5) $\boxed{\displaystyle\int \{f(x) - g(x)\}dx = \int f(x)dx - \int g(x)dx}$

$\because \dfrac{d}{dx}(\int f(x)\,dx - \int g(x)\,dx) = \dfrac{d}{dx}(\int f(x)\,dx + \int -g(x)\,dx)$ by 3)

$= \dfrac{d}{dx}(\int f(x)\,dx) + \dfrac{d}{dx}(\int -g(x)\,dx) = f(x) + (-g(x))$ by 4)

$= f(x) - g(x)$ $\therefore \int\{f(x) - g(x)\}\,dx = \int f(x)\,dx - \int g(x)\,dx$

Example   Integrate the expression:

(1) $\int dx = \int 1\, dx = x + C$

(2) $\int (-3)\, dx = -3x + C$

(3) $\int 0\, dx = C$

(4) $\int x\, dx = \int x^1\, dx = \frac{1}{1+1} x^{1+1} + C = \frac{1}{2} x^2 + C$

(5) $\int 4x^5\, dx = 4 \int x^5\, dx = 4\left(\frac{1}{5+1} x^{5+1} + C_1\right) = \frac{4}{6} x^6 + 4C_1 = \frac{2}{3} x^6 + C$

(6) $\int 2x\sqrt[3]{x}\, dx = 2 \int x^1 x^{\frac{1}{3}}\, dx = 2 \int x^{\frac{4}{3}}\, dx = 2\left(\frac{1}{\frac{4}{3}+1} x^{\frac{4}{3}+1} + C_1\right) = 2\left(\frac{3}{7} x^{\frac{7}{3}} + C_1\right)$

$$= \frac{6}{7} x^{\frac{7}{3}} + 2C_1 = \frac{6}{7}\left(x^2 \sqrt[3]{x}\right) + C$$

5-2 Integrating Trigonometric and Exponential Functions

1. The Antiderivatives of Trigonometric Functions

Note: $(\sin x)' = \cos x$

$(\cos x)' = -\sin x$

$(\tan x)' = \sec^2 x$

$(\cot x)' = -\csc^2 x$

$(\sec x)' = \sec x \tan x$

$(\csc x)' = -\csc x \cot x$

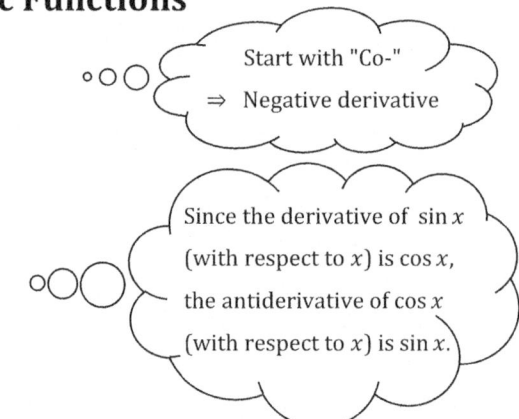

Start with "Co-"
⇒ Negative derivative

Since the derivative of $\sin x$ (with respect to x) is $\cos x$, the antiderivative of $\cos x$ (with respect to x) is $\sin x$.

Trig Antiderivatives

(1) $\displaystyle\int \cos x\, dx = \sin x + C$

(2) $\displaystyle\int \sin x\, dx = -\cos x + C$

(3) $\displaystyle\int \sec^2 x\, dx = \tan x + C$

(4) $\displaystyle\int \csc^2 x\, dx = -\cot x + C$

(5) $\displaystyle\int \sec x \tan x\, dx = \sec x + C$

(6) $\displaystyle\int \csc x \cot x\, dx = -\csc x + C$

(7) $\displaystyle\int \tan x\, dx = -\log|\cos x| + C$

(8) $\displaystyle\int \cot x\, dx = \log|\sin x| + C$

(9) $\displaystyle\int \sec x\, dx = \log|\sec x + \tan x| + C$

(10) $\displaystyle\int \csc x\, dx = \log|\csc x - \cot x| + C$

$$= -\log|\csc x + \cot x| + C$$

For example, $\displaystyle\int 2\cos x\,dx = 2\int \cos x\,dx = 2\sin x + C$

$$\int \tan^2 x\,dx = \int(\sec^2 x - 1)dx = \tan x - x + C$$

2. The Antiderivatives of Exponential Functions

Note: $\quad (e^x)' = e^x$

$$(a^x)' = a^x \log a \quad ; \quad \frac{(a^x)'}{\log a} = a^x \quad ; \quad \left(\frac{a^x}{\log a}\right)' = a^x$$

Exponential/Log Antiderivatives

$$(1)\ \int e^x dx = e^x + C$$

$$(2)\ \int a^x dx = \frac{a^x}{\log a} + C \quad (a > 0, a \neq 1)$$

$$(3)\ \int \log x\,dx = x\log x - x + C$$

For example, $\displaystyle\int (e^x - 3x + 4)dx = e^x - \frac{3}{2}x^2 + 4x + C$

$$\int e^{x+2}dx = \int(e^x e^2)dx = e^2\int e^x dx = e^2 e^x + C = e^{x+2} + C$$

$$\int(2e^x + 3^x)dx = 2\int e^x dx + \int 3^x dx = 2e^x + \frac{3^x}{\log 3} + C$$

5-3 Techniques of Integration

1. Integration by Substitution

(1) Form of $\int (ax + b)dx$

Consider the problem of finding $\displaystyle\int (2x + 1)^2 dx$.

You cannot apply the power rule $\displaystyle\int x^n dx = \frac{1}{n+1}x^{n+1} + C \ (n \neq 1)$ and obtain

$\displaystyle\int (2x + 1)^2 dx = \frac{1}{3}(2x + 1)^3 + C$ because $\displaystyle\int (2x + 1)^2 dx$ is not the form $\displaystyle\int x^2 dx$.

Let $2x + 1 = t$

Then, $x = \frac{t-1}{2}$ $\quad \therefore \frac{dx}{dt} = \frac{1}{2}$; $\quad dx = \frac{1}{2}dt$

$$\int (2x+1)^2 dx = \int (2x+1)^2 \frac{dx}{dt} dt = \int \left(t^2 \cdot \frac{1}{2}\right) dt = \frac{1}{2} \int t^2 dt = \frac{1}{2}\left(\frac{1}{3}t^3 + C_1\right) = \frac{1}{6}t^3 + C$$

For a differentiable function g, let $x = g(t)$ and $F = \int f(x)dx$.

Then, $\dfrac{dF}{dx} = f(x)$, $\dfrac{dx}{dt} = g'(t)$

$\therefore \dfrac{dF}{dt} = \dfrac{dF}{dx} \cdot \dfrac{dx}{dt} = f(x)g'(t) = f(g(t))g'(t)$

$\therefore F = \int f(g(t))g'(t)dt$.

Therefore, we have the following results:

$$\boxed{\text{If } x = g(t), \text{ then } \quad \int f(x)dx = \int \left(f(x) \cdot \frac{dx}{dt}\right) dt = \int f(g(t))g'(t)dt}$$

Example (1) $\displaystyle\int (2x-1)^3 dx$ (2) $\displaystyle\int e^{2x+1} dx$

(1) Let $2x - 1 = t$

Then, $2\dfrac{dx}{dt} = 1$; $\quad dx = \dfrac{1}{2}dt$

$\therefore \displaystyle\int (2x-1)^3 dx = \int t^3 \cdot \frac{1}{2} dt = \frac{1}{2} \int t^3 dt = \frac{1}{2} \cdot \frac{1}{4}t^4 + C = \frac{1}{8}t^4 + C = \frac{1}{8}(2x-1)^4 + C$

(2) Let $2x + 1 = t$

Then, $2\dfrac{dx}{dt} = 1$; $\quad dx = \dfrac{1}{2}dt$

$\displaystyle\int e^{2x+1} dx = \int e^t \cdot \frac{1}{2} dt = \frac{1}{2} \int e^t dt = \frac{1}{2}e^t + C$

Note: $\boxed{\displaystyle\int f(x)dx = F(x) + C \;\Rightarrow\; \int f(ax+b)dx = \frac{1}{a}F(ax+b) + C, \quad a \neq 0}$

$\because$ Let $ax + b = t$ Then, $a\dfrac{dx}{dt} = 1$; $\quad dx = \dfrac{1}{a}dt$

$\displaystyle\int f(ax+b)dx = \int f(t) \cdot \frac{1}{a} dt = \frac{1}{a}F(t) + C = \frac{1}{a}F(ax+b) + C, \quad a \neq 0$

Note: $\boxed{\begin{array}{l} \displaystyle\int \sin(ax+b)\,dx = -\frac{1}{a}\cos(ax+b) + C, \quad a \neq 0 \\[2mm] \displaystyle\int \cos(ax+b)\,dx = \frac{1}{a}\sin(ax+b) + C, \quad a \neq 0 \end{array}}$

(2) Form of $\int f(g(x))g'(x)dx$

When $F(x) = \int f(x)dx$ and g is differentiable function,

$$\frac{d}{dx}\{F(g(x)) + C\} = F'(g(x))g'(x) = f(g(x))g'(x)$$

Example

1) $\int (x^2 + 1)^3 \cdot 2x \, dx$

Let $x^2 + 1 = t$ Then, $2x\frac{dx}{dt} = 1$; $2xdx = dt$

$\therefore \int (x^2 + 1)^3 \cdot 2x \, dx = \int t^3 dt = \frac{1}{4}t^4 + C = \frac{1}{4}(x^2 + 1)^4 + C$

2) $\int (e^x + 1)^2 \cdot e^x \, dx$

Let $e^x + 1 = t$ Then, $e^x\frac{dx}{dt} = 1$; $e^x dx = dt$

$\therefore \int (e^x + 1)^2 \cdot e^x \, dx = \int t^2 dt = \frac{1}{3}t^3 + C = \frac{1}{3}(e^x + 1)^3 + C$

3) $\int (x^3 + 1)^2 \cdot x^2 \, dx = \int (x^3 + 1)^2 \cdot 3x^2 \cdot \frac{1}{3} dx = \frac{1}{3}\int (x^3 + 1)^2 \cdot 3x^2 dx$

$\qquad = \frac{1}{3}\int (x^3 + 1)^2 \cdot (x^3 + 1)' dx = \frac{1}{3} \cdot \frac{1}{3}(x^3 + 1)^3 + C$

$\qquad = \frac{1}{9}(x^3 + 1)^3 + C$

When $g(x) = t$, $\int f(g(x))g'(x) \, dx = \int f(t)dt$

---

(3) Form of $\int \frac{f'(x)}{f(x)} dx$

Note that $(\log x)' = \frac{1}{x}$, $x > 0$

$(\log|x|)' = \frac{1}{x}$, $x \neq 0$

$\because$ i) If $x > 0$, then $|x| = x$

$\qquad \therefore (\log|x|)' = (\log x)' = \frac{1}{x}$

ii) If $x < 0$, then $|x| = -x$

$\qquad \therefore (\log|x|)' = (\log(-x))' = \frac{1}{-x}(-x)' = \frac{1}{-x}(-1) = \frac{1}{x}$

Therefore, the corresponding integration formula for $(\log|x|)' = \frac{1}{x}$, $x \neq 0$, is

$$\int \frac{1}{x} \, dx = \log|x| + C, \qquad x \neq 0$$

Note: $\boxed{\displaystyle\int \frac{1}{ax+b} \, dx = \frac{1}{a}\log|ax+b| + C}$

Example

1) $\displaystyle\int \frac{2x}{x^2 + 1} \, dx$

    Let $x^2 + 1 = t$ Then, $2x\frac{dx}{dt} = 1$; $2x\,dx = dt$

$$\int \frac{2x}{x^2 + 1} \, dx = \int \frac{1}{t} \, dt = \log|t| + C = \log|x^2 + 1| + C = \log(x^2 + 1) + C \quad (\because x^2 + 1 > 0)$$

2) $\displaystyle\int \frac{e^x}{e^x + 1} \, dx$

    Let $e^x + 1 = t$ Then, $e^x\frac{dx}{dt} = 1$; $e^x\,dx = dt$

$$\int \frac{e^x}{e^x + 1} \, dx = \int \frac{1}{t} \, dt = \log|t| + C = \log|e^x + 1| + C = \log(e^x + 1) + C \quad (\because e^x + 1 > 0)$$

3) $\displaystyle\int \frac{3}{2x + 5} \, dx$

    Let $2x + 5 = t$ Then, $2\frac{dx}{dt} = 1$; $2\,dx = dt$

$$\int \frac{3}{2x + 5} \, dx = \frac{3}{2} \int \frac{2}{2x + 5} \, dx = \frac{3}{2} \int \frac{1}{t} \, dt = \frac{3}{2}\log|t| + C = \frac{3}{2}\log|2x + 5| + C$$

Note: $\boxed{\displaystyle\int \frac{f'(x)}{f(x)} \, dx = \log|f(x)| + C}$

(4) Integration by Partial Fractions

<u>Distinct Linear Factors:</u>

$$\boxed{\dfrac{px+q}{(ax+b)(cx+d)} = \dfrac{m}{ax+b} + \dfrac{n}{cx+d}}$$

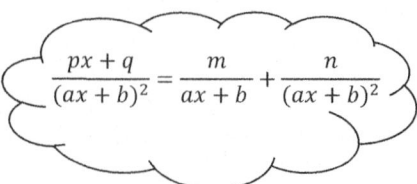

$$\frac{px + q}{(ax + b)^2} = \frac{m}{ax + b} + \frac{n}{(ax + b)^2}$$

Consider $\displaystyle\int \frac{1}{x^2 - 4}\, dx$.

1) Factor the denominator:

$$x^2 - 4 = (x + 2)(x - 2)$$

2) Write $\dfrac{1}{x^2-4} = \dfrac{a}{x+2} + \dfrac{b}{x-2}$ and clear of fraction to obtain:

$$\frac{1}{x^2-4} = \frac{a(x-2)+b(x+2)}{(x+2)(x-2)} = \frac{(a+b)x-2(a-b)}{(x+2)(x-2)}$$

3) Determine the constants. Equate coefficients of like powers of x and solve for the constants:

$$a + b = 0,\ \ a - b = -\frac{1}{2}$$

$$\therefore\ \ 2a = -\frac{1}{2}\ ;\ \ a = -\frac{1}{4}\ ,\ b = \frac{1}{4}$$

4) $\dfrac{1}{x^2-4} = \dfrac{\frac{1}{-4}}{x+2} + \dfrac{\frac{1}{4}}{x-2}$

$$\therefore\ \int \frac{1}{x^2 - 4}\, dx = \int \left(\frac{\frac{1}{-4}}{x + 2} + \frac{\frac{1}{4}}{x - 2} \right) dx = \frac{1}{-4}\int \frac{1}{x + 2}\, dx + \frac{1}{4}\int \frac{1}{x - 2}\, dx$$

$$= -\frac{1}{4}\log(x + 2) + \frac{1}{4}\log(x - 2) + C = \frac{1}{4}\log\left|\frac{x-2}{x+2}\right| + C$$

2. Integration by Parts

By the derivative of a product of two functions, we have $\{f(x)g(x)\}' = f'(x)g(x) + f(x)g'(x)$.

By integrating both members of this equation, we have

$$f(x)g(x) = \int f'(x)g(x)\, dx + \int f(x)g'(x)\, dx$$

$$\therefore\ \boxed{\int f'(x)g(x)\, dx = f(x)g(x) - \int f(x)g'(x)\, dx}$$

Let $f(x) = u$ and $g(x) = v$

Then, $\boxed{\displaystyle\int u'v = uv - \int uv'}$: Integration by parts (Indefinite integrals)

Example Find $\int x \cos x \, dx$

i) Let $u = x, \;\; v' = \cos x \, dx$

  Then, $u' = dx, \;\; v = \sin x$

  $\therefore \int x \cos x \, dx = x \sin x - \int \sin x \, dx = x \sin x - (-\cos x) + C = x \sin x + \cos x + C$

ii) Let $u = \cos x, \;\; v' = x dx$

  Then, $u' = -\sin x \, dx, \;\; v = \dfrac{1}{2}x^2$

  $\therefore \int x \cos x \, dx = \cos x \cdot \dfrac{1}{2}x^2 - \int \dfrac{1}{2}x^2(-\sin x)\,dx = \dfrac{1}{2}x^2 \cos x + \boxed{\dfrac{1}{2}\int x^2 \sin x \, dx}$

  This new integral is more complicated than the original one.
  Thus, we need a wise choice for u and v'.

Example Find $\int x \sin x \, dx$

Let $u = x, \;\; v' = \sin x \, dx$

Then, $u' = dx, \;\; v = -\cos x$

$\therefore \int \underset{u}{x} \; \underset{v'}{\sin x \, dx} = \underset{u}{x} \underset{v}{(-\cos x)} - \int \underset{v}{(-\cos x)} \underset{u'}{dx} = -x \cos x + \sin x + C$

Example Find $\int x e^x \, dx$

Let $u = x, \;\; v' = e^x dx$

Then, $u' = dx, \;\; v = e^x$

$\therefore \int \underset{u}{x} \; \underset{v'}{e^x dx} = \underset{u}{x} \; \underset{v}{e^x} - \int \underset{v}{e^x} \underset{u'}{dx} = xe^x - e^x + C$

Example Find $\int x \log x \, dx$

Let $u = \log x, \;\; v' = x dx$

Then, $u' = \dfrac{1}{x}dx, \;\; v = \dfrac{1}{2}x^2$

$\therefore \int \underset{u}{\log x} \; \underset{v'}{x dx} = \underset{u}{\log x} \underset{v}{\dfrac{1}{2}x^2} - \int \underset{v}{\dfrac{1}{2}x^2} \underset{u'}{\dfrac{1}{x}dx} = \dfrac{1}{2}x^2 \log x - \dfrac{1}{2}\int x \, dx = \dfrac{1}{2}x^2 \log x - \dfrac{1}{4}x^2 + C$

Exercises

#1 Find the values of the constants $a, b,$ and c.

(1) $\int (3x^2 + ax - 4)dx = bx^3 + 2x^2 + cx + 3$

(2) $\frac{d}{dx}[\int(ax^2 + 3x + 2)dx] = 8x^2 + bx + c$

#2 Integrate the expression.

(1) $\int x^5 dx$

(2) $\int (3x^2 + 6x)dx$

(3) $\int (2x^2 + 3x - 4)dx$

(4) $\int \left(\frac{2}{x} - \frac{3}{x^2}\right) dx$

(5) $\int \left(\frac{2x^3 - x^4}{x^4}\right) dx$

(6) $\int x^2 (3 - \sqrt{x})dx$

(7) $\int x(x - 1)(x + 2)dx$

(8) $\int (\sin\theta + \cos\theta)^2 d\theta + \int (\sin\theta - \cos\theta)^2 d\theta$

(9) $\int \left(\frac{\sin^2 x}{\cos^2 x} - \frac{1}{\cos^2 x}\right) dx$

(10) $\int \frac{x^3}{x+1} dx + \int \frac{1}{x+1} dx$

(11) $\int \left(\frac{x^2 - 2x - 3}{\sqrt{x}}\right) dx$

(12) $\int \left(x - \frac{1}{x}\right)^3 dx$

(13) $\int \frac{(\sqrt{x}-2)^2}{x} dx$

(14) $\int \left(\frac{x^4 + x^2 + 1}{x^2 - x + 1}\right) dx$

#3 For two functions $f(x)$ and $g(x)$ satisfying $\frac{d}{dx}\{f(x) + g(x)\} = 4$ and $\frac{d}{dx}\{f(x)g(x)\} = 8x$, $f(0) = -1$ and $g(0) = 1$. Find the following functions:

(1) $f(x) + g(x)$

(2) $f(x)g(x)$

(3) $f(x), \ g(x)$

#4 Answer the question.

(1) When $f'(x) = 2x^2 - 3x + 1$ and $f(1) = 2$, find the function $f(x)$.

(2) For the problem to integrate $f(x)$, you misunderstand to differentiate it and obtain $3x + 4$. Integrate $f(x)$ and find the coefficient of x^2 in the integral.

(3) When $f(x)$ is the antiderivative of $x^2\sqrt{x} - x + 3$ such that $f(0) = 4$, find $f(x)$.

(4) Find the solution of the equation $\log_x \left\{\frac{d}{dx}(\int x^2 dx)\right\} = 2x^3 + 3x^2 - 8x + 5$.

(5) For the graph of a function $y = f(x)$ passing through the two points $(1, 0)$ and $(-1, 1)$, $f''(x) = 2x - 4$. Find the sum of all solutions of $f(x) = 0$.

(6) When the slope of the tangent line at (a, b) on the graph of $y = f(x)$ is $\frac{1}{a}$ $(a > 0)$, find the function passing through the point $(1, 3)$.

(7) For a continuous function $f(x)$ satisfying $f'(x) = |x|$ for any real number x, $f(1) = 2$.

Find the value of $f(-1)$.

(8) For a continuous function $f(x)$, the derivative of it is $f'(x) = \begin{cases} 2x, & x < 1 \\ 1, & x \geq 1 \end{cases}$. When the graph

of the function $y = f(x)$ passes through a point $(0, 1)$, find the value of $f(10) + f(-10)$.

(9) When $\int (1 - f(x))dx = \dfrac{1}{2}x^2(4 - x^2) + C$, find the local extreme values of $f(x)$.

(10) For a function $y = f(x)$ with a local extreme value 4 at $x = 1$, $f'(x) = 3x^2 - 8x + a$.

Find the values of a and the other local extreme value.

(11) When $f'(x) = x^2 - 4x + 3$ and the local maximum value of $f(x)$ is 5, find the function

$f(x)$ and the local minimum value of $f(x)$.

(12) For a function $f(x)$ satisfying $f'(x) = x^2 - x - 6$, find the difference between the local

extreme values.

(13) For a function $f(x)$, the derivative of $f(x)$ is shown

as the Figure. When the local maximum value of $f(x)$ is 4 and

local minimum value of $f(x)$ is -2, find the value of $f(0)$.

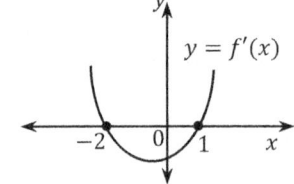

(14) For a function $f(x)$, the derivative of $f(x)$ is shown

as the Figure. When $f(0) = 1$, find the range of k so that

the equation $f(x) = kx + 1$ of x has three different

real number solutions.

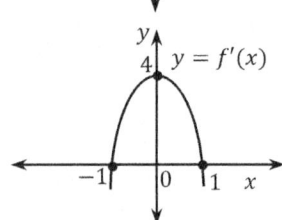

(15) When $f(x) = \int (x \log x + e^x + x)dx$, find the limit $\displaystyle\lim_{h \to 0} \dfrac{f(1-2h) - f(1)}{h}$.

(16) For a cubic function $f(x)$, the antiderivative of $f(x)$ is $F(x) = xf(x) - 2x^3(x - 1)$.

When $f(-1) = 0$, find the value of $f(1)$.

(17) For a function $f(x) = x \log x - x$, $g(x)$ is the inverse of $f'(x)$.

When the antiderivative of $g(x)$ at $x = 1$ is $3e$, find the antiderivative of $g(x)$.

(18) For two polynomial functions $f(x)$ and $g(x)$,

(i) $f(x) + g(x) = 2x^3 - 2x^2 + 2x + 4$ and

(ii) $f'(x) - g'(x) = 6x^2 - 4$.

When the two graphs of the functions intersect at $x = 1$, find the value of $f(1) - g(-1)$.

(19) For a function $f(x) = 1 + 2x + 3x^2 + \cdots\cdots + nx^{n-1}$, the indefinite integral $F(x)$ satisfies

$\displaystyle\lim_{n \to \infty} F\left(\dfrac{1}{2}\right) = 10$. Find the value of $F(0)$.

#5   Find the indefinite integrals.

(1) $\int (3 - \sin x)dx$

(2) $\int \cot^2 x\, dx$

(3) $\int \frac{1}{1-\sin^2 x} dx$

(4) $\int \frac{\sin^2 x}{1-\cos x} dx$

(5) $\int \frac{\sin x}{\cos^2 x} dx$

(6) $\int \frac{\cos x}{\sin^2 x} dx$

(7) $\int \frac{\sin x + \cos x}{\sin x \cdot \cos x} dx$

(8) $\int \frac{4+\sin x}{\cos x} dx$

(9) $\int \sin^2 \frac{x}{2} dx$

(10) $\int (\tan x + \cot x)^2 dx$

(11) $\int \frac{2e^x \sin^2 x - 3}{\sin^2 x} dx$

(12) $\int \frac{8^x + 1}{2^x + 1} dx$

(13) $\int \frac{x - \cos^2 x}{x \cos^2 x} dx$

#6   Answer the question.

(1) When $f'(x) = \frac{1}{\tan \frac{x}{2} + \cot \frac{x}{2}}$ and $f\left(\frac{\pi}{2}\right) = -1$, find the function $f(x)$. $(0 < x < \pi)$

(2) When the slope of the tangent line at (x, y) on the graph of $y = f(x)$ is at a rate proportional to $e^x + x$ and the equation of the tangent line at $x = 0$ on the graph is $y = 2x + 1$, find the value of $f(1)$.

(3) At every point of a curve y, $y'' = x^2 - 1$. When the equation of the tangent line at $(1, -1)$ on the curve y is $x + 2y = 3$, find the equation of the curve y.

(4) Find the equation of the curve passing through the point $(0, 2)$. The slope of the curve at any point $P(x, y)$ is $= 3x^2 y$.

#7   A continuous function $f(x)$ on $(-\infty, \infty)$ satisfies $f(0) = 1$ and $f'(x) = -f(x) + e^{-x} \sin x$.

(1) Find the derivative of $e^x f(x)$.

(2) Find the function $f(x)$.

#8   Find the antiderivative by performing the variable substitution.

(1) $\int (x^3 + 2)^2 \cdot 3x^2 dx$

(2) $\int (x^3 + 2)^{\frac{1}{2}} \cdot x^2 dx$

(3) $\int (x^2 + 2x + 1)^3 (x + 1)dx$

(4) $\int \frac{8x^2}{(x^3+4)^3} dx$

(5) $\int \frac{x^2}{\sqrt[4]{x^3+1}} dx$

(6) $\int 2x\sqrt{1 - 3x^2} dx$

(7) $\int x\sqrt{x^2 + 1} dx$

(8) $\int \frac{x-1}{\sqrt{x+1}} dx$

(9) $\int \frac{x+4}{\sqrt[3]{x^2+8x}} dx$

(10) $\int \sin x° dx$

(11) $\int \sin x \cos x \, dx$

(12) $\int \sin x \cos 2x \, dx$

(13) $\int \sin^2 x \cos x \, dx$

(14) $\int (\sin x + \cos x)^2 \, dx$

(15) $\int \cos^2 x \, dx$

(16) $\int \sin^2 x \, dx$

(17) $\int \sin 2x \cos 3x \, dx$

(18) $\int \cos 4x \cos 2x \, dx$

(19) $\int (1 + \cos x)^4 \sin x \, dx$

(20) $\int \frac{\sin x}{\cos x} \, dx$

(21) $\int \frac{\sec^2 x}{\tan x} \, dx$

(22) $\int \frac{1}{1+\cos x} \, dx$

(23) $\int \frac{1}{\sin x} \, dx$

(24) $\int \frac{\sin \sqrt{x}}{\sqrt{x}} \, dx$

(25) $\int (e^x + 1)^4 \cdot e^x \, dx$

(26) $\int e^{-4x} \, dx$

(27) $\int x e^{x^2} \, dx$

(28) $\int \frac{e^x - 1}{e^x + 1} \, dx$

(29) $\int e^x \cos e^x \, dx$

(30) $\int e^{3 \cos 2x} \sin 2x \, dx$

(31) $\int a^{2x} \, dx$

(32) $\int 5^{4x+3} \, dx$

#9 Find the integrals.

(1) $\int \frac{1}{x+2} \, dx$

(2) $\int \frac{x+2}{x+1} \, dx$

(3) $\int \frac{x^2+1}{x-1} \, dx$

(4) $\int \frac{x}{x^2-1} \, dx$

(5) $\int \frac{x^2}{1-2x^3} \, dx$

(6) $\int \frac{2x+1}{x^2+x+1} \, dx$

(7) $\int \frac{3}{2x^2+x-1} \, dx$

(8) $\int \frac{1}{25-16y^2} \, dy$

(9) $\int \frac{1}{9x^2-16} \, dx$

(10) $\int \frac{x+1}{(x-1)(x-2)} \, dx$

(11) $\int \frac{1}{\sqrt{x+1}+\sqrt{x}} \, dx$

(12) $\int \frac{1}{\sqrt[3]{2x+1}} \, dx$

(13) $\int \tan x \, dx$

(14) $\int \frac{1+\cos x}{x+\sin x} \, dx$

(15) $\int \frac{1}{x \log x} \, dx$

(16) $\int \frac{1}{e^x+1} \, dx$

#10 Use integration by parts to perform the indicated integration.

(1) $\int x e^{3x} \, dx$

(2) $\int x^2 e^{2x} \, dx$

(3) $\int x \cos 2x \, dx$

(4) $\int x^2 \sin x \, dx$

(5) $\int \sin^2 x \, dx$

(6) $\int x^2 \log x \, dx$

(7) $\int \log(x + 2) \, dx$

(8) $\int x \sqrt{1 + x} \, dx$

(9) $\int \frac{\log(x-1)}{\sqrt{x-1}} \, dx$

#11 Find the value.

(1) For a function $f(x) = \int e^x \cos x \, dx$ satisfying $f(0) = \frac{1}{2}$,

    1) Find $f(x)$.

    2) Find the value of x such that $f(x) = \frac{1}{2} e^x$ $(0 < x < 2\pi)$

(2) When the slope of the tangent line at (x, y) on the graph of $y = f(x)$ is $x \log x$,

    find $f(x)$ passing through a point $(1, 3)$.

(3) For two differentiable functions $f(x)$ and $g(x)$, $f(x) > 0$ and $f(0) = e^2$.

    Find $f(x)$ such that $f(x)\{g(x) + g'(x)\} = (f(x)g(x))'$. $(g(x) \neq 0)$

(4) For a function $f(x)$ defined on real numbers, $f'(x) = \frac{3x}{x^2+1}$ and $f(0) = 1$.

    Find the value of $f(\sqrt{e-1})$.

(5) For a function $f(x) = \int \frac{\log x}{x} \, dx$, $f(1) = -1$. Find the value of x such that $f(x) = \log x - 1$.

(6) When a function $f(x)$ defined on an open interval $\left(0, \frac{3\pi}{4}\right)$ satisfies:

    i) $f'(x) = \sin 2x - \cos x$ and

    ii) $f(x)$ has a local minimum value 1,

    find the difference between the local extreme values.

(7) Let $F(x) = \int f(x) dx$ $(x > 0)$.

    When $F(x) = xf(x) - x^2 \sin x$ and $F\left(\frac{\pi}{2}\right) = \frac{\pi}{4}$, find the value of $f(\pi)$.

(8) The rate of change of y with respect to x is $2x^3$, and $y = 12$ when $x = 2$.

    Find the value of y when $x = -2$.

(9) A certain quantity a increases at a rate proportional to itself. If $a = 8$ when $t = 0$ and $a = $

    32 when $t = 2$, find the value of a when $t = 4$.

(10) A ball is rolled over a level lawn with initial velocity 30 feet/sec. Due to friction, the

    velocity decreases at the rate of 4 feet/sec^2. How far will the ball roll?

(11) A ball dropped from a balloon 648 feet above the ground.

    When the balloon was rising at the rate of 36 feet/sec, find:

    1) The greatest distance above the ground attained by the ball.

    2) The time the ball was in the air.

    3) The speed of the ball when it struck the ground.

    Assume positive distanced velocity to be directed upward.

    $a = \frac{dV}{dt} = -24$ feet/sec^2, and $V = -24t + C$

Chapter 6. The Definite Integral

6-1 Integral and Area

The two main ideas in Calculus are developed from geometric ideas related to curves. The derivative comes from the tangent line to a curve and the definite integral comes from the area of a curved region.

1. Defining the Area of a Region

The area of a rectangle is its length times its width;

$$S = ab$$

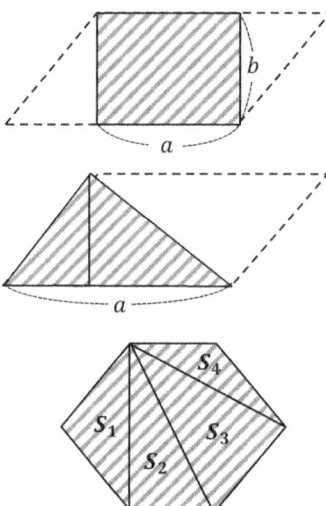

The area of a triangle is one-half the base times the altitude;

$$S = \frac{1}{2}ah$$

Since any polygon can be broken up into triangles, we derive the formulas for the area of a polygon;

$$S = S_1 + S_2 + S_3 + S_4$$

(1) Area of a circle

Consider a region with a curved boundary.

The problem of calculating the area is more difficult.

The approximation should get better and better as n gets larger.

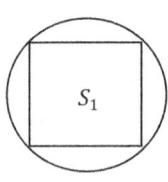

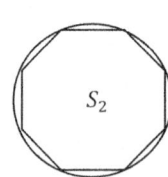

4 sides	8 sides	12 sides

Consider regular inscribed polygons $S_1, S_2,$ and S_3 with 4 sides, 8 sides, and 12 sides as shown in Figure.

The area of inscribed polygon S_n with n sides is

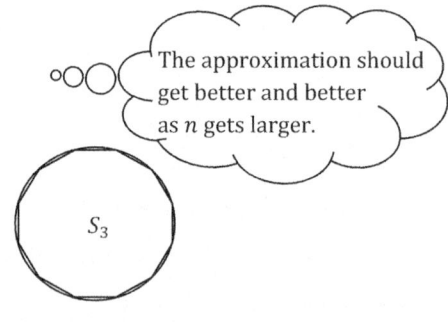

$$S_n = (\text{the area of } \Delta\, OAB) \times n = \left(\frac{1}{2} \cdot \overline{AB} \cdot h\right) \cdot n = \frac{1}{2}hn\overline{AB}$$

Let the perimeter of the polygon S_n be $\overline{AB} \cdot n = l_n$ Then, $S_n = \frac{1}{2}hl_n$

For very large n, an inscribed side $\overline{AB}$ becomes indistinguishable from the arc $\overparen{AB}$ and the area of the inscribed polygon differs from the area of the circle by a very small amount.

Therefore, the area of the circle is the limiting value of the area of an inscribed regular polygon of n sides as the number of sides n is increased indefinitely.

Since $l_n \to 2\pi r$ and $h \to r$ as $n \to \infty$,

the area of the circle is $S = \lim\limits_{n\to\infty} S_n = \lim\limits_{n\to\infty} \frac{1}{2} h l_n = \frac{1}{2} \cdot r \cdot 2\pi r = \pi r^2$

(2) Area under $y = x^2$

1) Area by Circumscribed Polygon

Consider the area S bounded by the curve $y = x^2$, the x-axis$(y = 0)$, and the vertical line $x = 1$

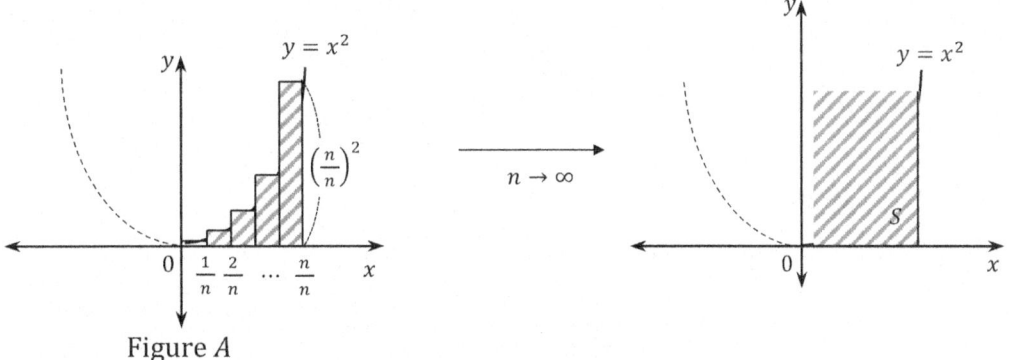

Figure A

Divide the interval $[0, 1]$ up into n equal subintervals by the points

$$x_0 = 0, \; x_1 = \frac{1}{n}, \; x_2 = \frac{2}{n}, \; x_3 = \frac{3}{n}, \; \cdots\cdots, \; x_n = \frac{n}{n} = 1$$

The area S is smaller than the combined area S_n of the rectangles formed as in Figure A.

That is, $S_n = \frac{1}{n}\left(\frac{1}{n}\right)^2 + \frac{1}{n}\left(\frac{2}{n}\right)^2 + \frac{1}{n}\left(\frac{3}{n}\right)^2 + \cdots\cdots + \frac{1}{n}\left(\frac{n}{n}\right)^2 > S$

or $\dfrac{n(n+1)(2n+1)}{6n^3} > S$

$S = \lim\limits_{n\to\infty} S_n = \lim\limits_{n\to\infty} \dfrac{n(n+1)(2n+1)}{6n^3} = \dfrac{2}{6} = \dfrac{1}{3}$

2) Area by Inscribed Polygon

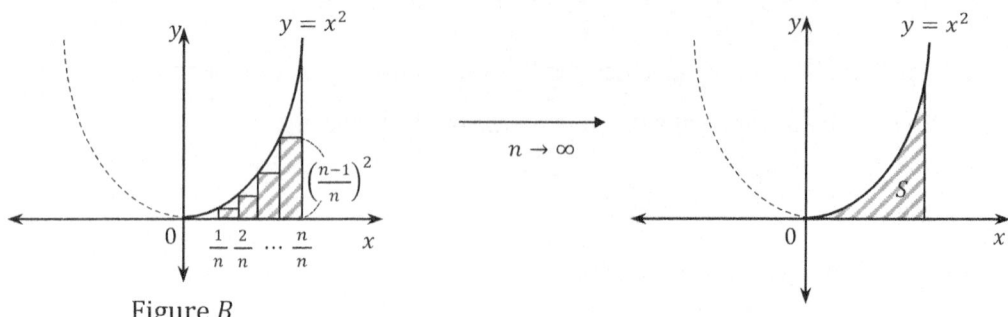

Figure B

The area S is larger than the combined area $S_n{}'$ of the rectangles formed as in Figure B.

That is, $S_n{}' = \frac{1}{n}\left(\frac{1}{n}\right)^2 + \frac{1}{n}\left(\frac{2}{n}\right)^2 + \frac{1}{n}\left(\frac{3}{n}\right)^2 + \cdots\cdots + \frac{1}{n}\left(\frac{n-1}{n}\right)^2 < S$

or $\quad \frac{(n-1)n(2n-1)}{6n^3} < S$

$S = \lim\limits_{n\to\infty} S_n{}' = \lim\limits_{n\to\infty} \frac{(n-1)n(2n-1)}{6n^3} = \frac{2}{6} = \frac{1}{3}$

Therefore, we conclude that $S = \lim\limits_{n\to\infty} S_n = \lim\limits_{n\to\infty} S_n{}'$

(3) Area under $y = x^n$

Consider the area bounded by the curve $y = x^3$, the x-axis $(y = 0)$, and the vertical line $x = 1$.

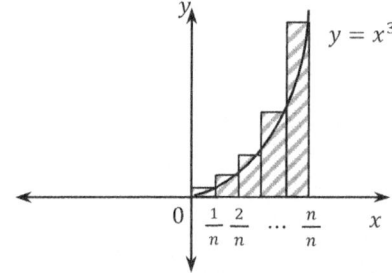

Divide the interval $[0, 1]$ up into n equal subintervals by the points:

$$x_0 = 0, \ x_1 = \frac{1}{n}, \ x_2 = \frac{2}{n}, \ \cdots\cdots, \ x_n = \frac{n}{n} = 1$$

Consider the rectangle with base $[\, x_{i-1,} \, x_i \,]$ and height $y = f(x_i) = x_i{}^3$.

The areas of the rectangles are $\frac{1}{n}\left(\frac{1}{n}\right)^3, \ \frac{1}{n}\left(\frac{2}{n}\right)^3, \ \frac{1}{n}\left(\frac{3}{n}\right)^3, \ \cdots\cdots$

The area S under the curve is

$$S = \lim\limits_{n\to\infty}\left\{\left(\frac{1}{n}\right)^3 + \left(\frac{2}{n}\right)^3 + \cdots\cdots + \left(\frac{n}{n}\right)^3\right\}\cdot\frac{1}{n} = \lim\limits_{n\to\infty}\frac{1}{n^4}(1^3 + 2^3 + \cdots\cdots + n^3)$$

$$= \lim\limits_{n\to\infty}\frac{1}{n^4}\left(\frac{n(n+1)}{2}\right)^2 = \lim\limits_{n\to\infty}\frac{n^2(n+1)^2}{4n^4} = \lim\limits_{n\to\infty}\frac{(n+1)^2}{4n^2} = \lim\limits_{n\to\infty}\frac{1}{4}\left(1 + \frac{2}{n} + \frac{1}{n^2}\right) = \frac{1}{4}$$

In a similar way, the area bounded by the curve $y = x^4$, the x-axis $(y = 0)$, and the vertical line $x = 1$ is

$$S = \lim_{n \to \infty} \left\{ \left(\frac{1}{n}\right)^4 + \left(\frac{2}{n}\right)^4 + \cdots\cdots + \left(\frac{n}{n}\right)^4 \right\} \cdot \frac{1}{n} = \lim_{n \to \infty} \frac{1}{n^5}(1^4 + 2^4 + \cdots\cdots + n^4)$$

$$= \lim_{n \to \infty} \frac{1}{n^5} \left(\frac{6n^5 + 15n^4 + 10n^3 - n}{30}\right) = \frac{6}{30} = \frac{1}{5}$$

Therefore, we see the pattern:

Bounded by curves	Area bounded
$y = x$, x-axis $(y = 0)$, $x = 1$	$\frac{1}{2}$
$y = x^2$, x-axis $(y = 0)$, $x = 1$	$\frac{1}{3}$
$y = x^3$, x-axis $(y = 0)$, $x = 1$	$\frac{1}{4}$
$y = x^4$, x-axis $(y = 0)$, $x = 1$	$\frac{1}{5}$

The area bounded by the curve $y = x^n$, the x-axis $(y = 0)$, and the vertical line $x = 1$ is

$\dfrac{1}{n+1}$ for $n = 0, 1, 2, \cdots\cdots$

Now, consider the area bounded by the curve $y = x^2$, the x-axis $(y = 0)$, and a line $x = b$.

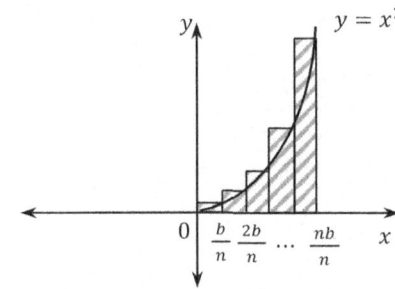

Divide the interval $[0, b]$ up into n equal subintervals by the points:

$$x_0 = 0, \quad x_1 = \frac{1}{n}b, \quad x_2 = \frac{2}{n}b, \quad \cdots\cdots, \quad x_n = \frac{n}{n}b = b$$

The area S under the curve is

$$S = \lim_{n \to \infty} \left\{ \left(\frac{1}{n} \cdot b\right)^2 \cdot \frac{b}{n} + \left(\frac{2}{n} \cdot b\right)^2 \cdot \frac{b}{n} + \cdots\cdots + \left(\frac{n}{n} \cdot b\right)^2 \cdot \frac{b}{n} \right\} = \lim_{n \to \infty} \frac{b^3}{n^3}(1^2 + 2^2 + \cdots\cdots + n^2)$$

$$= \lim_{n \to \infty} \frac{b^3}{n^3} \left(\frac{n(n+1)(2n+1)}{6}\right) = \frac{2b^3}{6} = \frac{1}{3}b^3$$

Similarly, the area bounded by the curve $y = x^3$, the x-axis $(y = 0)$, and a line $x = b$ is $\frac{1}{4}b^4$.

Now, consider the area bounded by the curve $y = x^2$, the x-axis, $x = a$, and $x = b$ $(a < b)$.

Since the area from 0 to b is $\frac{1}{3}b^3$ and the area from 0 to a is $\frac{1}{3}a^3$,

the area from a to b is $\frac{1}{3}b^3 - \frac{1}{3}a^3 = \frac{1}{3}(b^3 - a^3)$.

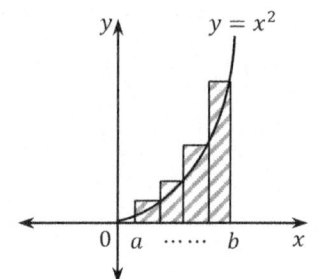

Therefore, we see the pattern:

Bounded by curves	Area bounded
$y = x$, x-axis $(y = 0)$, $x = a$, $x = b$	$\frac{1}{2}(b^2 - a^2)$
$y = x^2$, x-axis $(y = 0)$, $x = a$, $x = b$	$\frac{1}{3}(b^3 - a^3)$
$y = x^3$, x-axis $(y = 0)$, $x = a$, $x = b$	$\frac{1}{4}(b^4 - a^4)$
$y = x^4$, x-axis $(y = 0)$, $x = a$, $x = b$	$\frac{1}{5}(b^5 - a^5)$
$\vdots$	$\vdots$
$y = x^n$, x-axis $(y = 0)$, $x = a$, $x = b$	$\frac{1}{n+1}(b^{n+1} - a^{n+1})$

6-2 The Definite Integral

1. Riemann Sums

Consider the area bounded by a curve $y = f(x)$, the x-axis, $x = a$, and $x = b$ $(a < b)$.

Divide the interval $[a, b]$ up into n equal subintervals by the points:

$x_0 = a, \ x_1, \ x_2, \ \cdots\cdots, \ x_n = b$

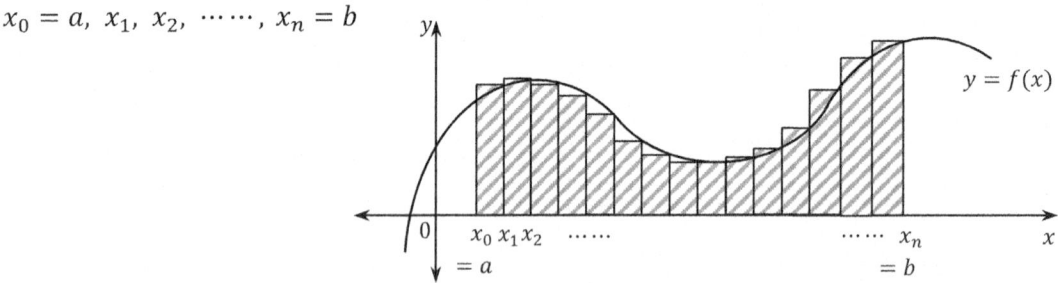

Let $\Delta x = \frac{b-a}{n}$ be the common interval.

Use x_1 to determine the height $f(x_1)$ of a rectangle, use x_2 to determine the height $f(x_2)$ of the next rectangle, etc. Then, the area of each rectangle is $f(x_i)\Delta x$.

Thus, the area of all rectangles under the curve from a to b is

$$f(x_1)\Delta x + f(x_2)\Delta x + \cdots\cdots + f(x_n)\Delta x = \{f(x_1) + f(x_2) + \cdots\cdots + f(x_n)\}\Delta x = \sum_{i=1}^{n} f(x_i)\Delta x$$

Let $\Delta x_i = x_i - x_{i-1}$

Choose an arbitrary point $\bar{x}_i$ (which may be a middle point or an end point), called a *sample point* for the ith subinterval.

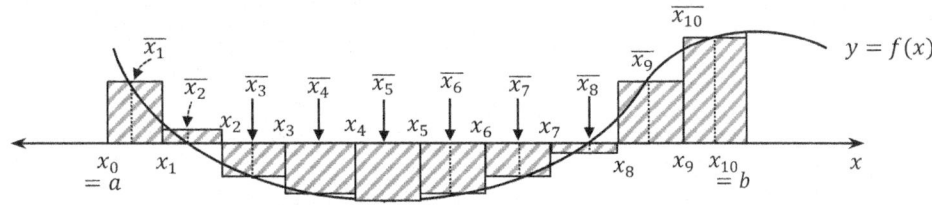

Then, the sum $S_P = \displaystyle\sum_{i=1}^{n} f(\bar{x}_i)\Delta x_i = S_1 + S_2 + (-S_3) + (-S_4) + (-S_5) + \cdots + S_9 + S_{10}$

We call S_P a *Riemann Sum* for f corresponding to the polygon S_i.

> The rectangle below the x-axis has the negative of its area.
> ($\because f(\bar{x}_i) < 0$)

2. The Definite Integral

(1) Definition

Let f be a positive, continuous function

on the closed interval $[a, b]$.

If $\displaystyle\lim_{n\to\infty}\sum_{i=1}^{n} f(x_i)\Delta x \left(\Delta x = \dfrac{b-a}{n}\right)$ exists,

we say f is integrable on $[a, b]$ and

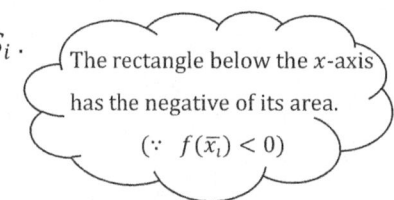

$\displaystyle\int_a^b f(x)dx = \lim_{n\to\infty}\sum_{i=1}^{n} f(x_i)\Delta x$ is called the *definite integral* or *Riemann integral* of f from a to b.

$\displaystyle\int_a^b f(x)dx = \lim_{n\to\infty}\sum_{i=0}^{n-1} f(x_i)\Delta x = \lim_{n\to\infty}\sum_{i=1}^{n} f(x_{i-1})\Delta x$

> The shaded area of the region is equal to the definite integral.

If $f(x) \geq 0$, then $S_n \geq 0$

$\quad f(x) < 0$, then $S_n < 0$

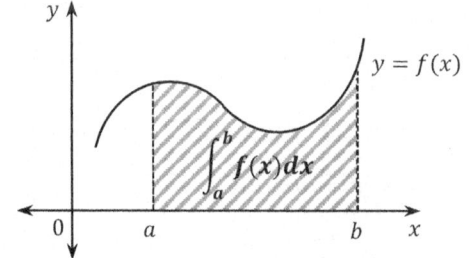

$\displaystyle\int_a^c f(x)dx$ is the signed area of the region trapped between the curve $y = f(x)$ and the x-axis

on the interval $[a, c]$.

That is, $\displaystyle\int_a^b f(x)dx = S_1, \qquad \int_b^c f(x)dx = -S_2 \qquad \therefore \int_a^c f(x)dx = S_1 - S_2$

Note: For a continuous function $f(x)$,

$$(1)\ \lim_{n \to \infty} \int_a^b f(x)dx = \lim_{n \to \infty} \sum_{i=1}^{n} f(x_i) \cdot \frac{b-a}{n}$$

$$(2)\ \lim_{n \to \infty} \sum_{i=1}^{n} f\left(a + \frac{(b-a)i}{n}\right) \cdot \frac{b-a}{n} = \int_a^b f(x)dx$$

(3) Letting $b - a = p$ (i.e., $b = a + p$), $\quad \lim_{n \to \infty} \sum_{i=1}^{n} f\left(a + \frac{pi}{n}\right) \cdot \frac{p}{n} = \int_a^{a+p} f(x)dx$

$\quad$ Specially, $\quad$ when $a = 0$, $\quad \lim_{n \to \infty} \sum_{i=1}^{n} f\left(\frac{pi}{n}\right) \cdot \frac{p}{n} = \int_0^p f(x)dx$

(4) Let $x = a + pt \quad$ Then, $dx = pdt$

$\quad t = 0$ when $x = a$

$\quad t = 1$ when $x = a + p$

$$\therefore\ \int_a^{a+p} f(x)dx = \int_0^1 f(a + pt)pdt = p\int_0^1 f(a+pt)dt = p\int_0^1 f(a+px)dx$$

$$\therefore\ \lim_{n \to \infty} \sum_{i=1}^{n} f\left(a + \frac{pi}{n}\right) \cdot \frac{1}{n} = \int_0^1 f(a+px)dx$$

For the symbol $\int_a^b f(x)dx$, the function f is called the *integrand* and $[a, b]$ is the *interval of integration.* The number a is the *lower limit (lower endpoint)* and b is the *upper limit (upper endpoint)* of integration.

$\int f(x)dx$ is called the *indefinite integral* of f. It is a function of x with an arbitrary constant C.

$\int_a^b f(x)dx$ is called the *definite integral* of f. It is a real number.

$$\int_a^b f(x)dx = \left[\int f(x)dx\right]_a^b$$

Example Evaluate the limit:

$(1)\ \displaystyle\lim_{n \to \infty} \sum_{i=1}^{n} \left(\frac{2i}{n}\right)^3 \frac{1}{n} = \int_0^1 (2x)^3 dx = 8\int_0^1 x^3 dx = 8\left[\frac{1}{4}x^4\right]_0^1 = 8\frac{1}{4}(1^4 - 0) = 2$

$(2)\ \displaystyle\lim_{n \to \infty} \sum_{j=1}^{n} \left(1 + \frac{j}{n}\right)^3 \frac{1}{n} = \int_0^1 (1+x)^3 dx = \frac{1}{4}\left[(1+x)^4\right]_0^1 = \frac{1}{4}(2^4 - 1) = \frac{15}{4}$

$(3)\ \displaystyle\lim_{n \to \infty} \sum_{k=0}^{n-1} \left(1 + \frac{2k}{n}\right)^3 \frac{3}{n} = \int_0^1 (1+2x)^3 3dx = 3\left[\frac{1}{4}(1+2x)^4 \cdot \frac{1}{2}\right]_0^1 = \frac{3}{8}(3^4 - 1) = 30$

$(4)\ \displaystyle\lim_{n \to \infty} \sum_{i=1}^{n} \left(1 + \frac{2i}{n}\right)^2 \frac{3}{n} = \int_0^1 (1+2x)^2 3dx = 3\left[\frac{1}{3}(1+2x)^3 \cdot \frac{1}{2}\right]_0^1 = \frac{1}{2}(3^3 - 1) = 13$

(2) The Fundamental Theorem of Integral

$$\int_a^b g'(x)dx = g(b) - g(a)$$

The Fundamental Theorem

Let f be continuous (hence integrable) on a closed interval $[a, b]$ and

F be any antiderivative of f. That is, $\int f(x)dx = F(x) + C$

Then, $\int_a^b f(x)dx = [F(x)]_a^b = F(b) - F(a)$

For example, evaluate $\int_1^2 x^3 dx$.

Find the antiderivative of the integrand x^3 using the power rule for integration. That is,

$$\int x^3 dx = \frac{1}{4}x^4 + C$$

Write the antiderivative with vertical bars to its left and right sides:

$$\int_1^2 x^3 dx = \left[\frac{1}{4}x^4 + C\right]_1^2$$

and evaluate the antiderivative at each boundary and then calculate the difference.

$$\int_1^2 x^3 dx = \left[\frac{1}{4}x^4 + C\right]_1^2 = \left(\frac{1}{4} \cdot 2^4 + C\right) - \left(\frac{1}{4} \cdot 1^4 + C\right) = 4 - \frac{1}{4} = \frac{15}{4}$$

Note: Since C of the indefinite integration always cancels out, we write the antiderivative

excluding C; i. e., $\int_1^2 x^3 dx = \left[\frac{1}{4}x^4\right]_1^2$

Example

(1) $\int_1^2 \frac{1}{x}dx = [\log|x|]_1^2 = \log|2| - \log|1| = \log 2$

(2) $\int_0^\pi \sin x\, dx = [-\cos x]_0^\pi = -\cos \pi - (-\cos 0) = 1 + 1 = 2$

(3) $\int_0^{\frac{\pi}{2}} \sin^2 x \cos x\, dx$

Let $\sin x = t$ Then, $\cos x\, dx = dt$

$$\int \sin^2 x \cos x\, dx = \int t^2 dt = \frac{1}{3}t^3 + C = \frac{1}{3}\sin^3 x + C$$

$$\int_0^{\frac{\pi}{2}} \sin^2 x \cos x\, dx = \left[\frac{1}{3}\sin^3 x\right]_0^{\frac{\pi}{2}} = \frac{1}{3}\sin^3 \frac{\pi}{2} - \frac{1}{3}\sin^3 0 = \frac{1}{3} - 0 = \frac{1}{3}$$

6-3 Properties of the Definite Integral

From the definition of $\displaystyle\int_a^b f(x)dx,$ we assumed that $a < b.$

---

(1) If $a = b$, then $\displaystyle\int_a^a f(x)dx = 0$

(2) If $a > b$, then $\displaystyle\int_a^b f(x)dx = -\int_b^a f(x)dx$

For example, $\displaystyle\int_0^0 f(x)dx = 0$; $\displaystyle\int_2^1 f(x)dx = -\int_1^2 f(x)dx$

Note: When $a > b$,

$$\int_a^b f(x)dx = -\int_b^a f(x)dx = -[F(x)]_b^a \qquad (\because \int f(x)dx = F(x) + C)$$

$$= -\{F(a) - F(b)\} = F(b) - F(a)$$

$$\therefore \ \text{If} \int f(x)dx = F(x) + C, \ \text{then} \int_a^b f(x)dx = [F(x)]_a^b = F(b) - F(a)$$

1. Properties of the Definite Integral

(1) Linearity of the Definite Integral

Suppose that f and g are integrable on $[a, b]$ and k is a constant.

Then, kf and $f + g$ are integrable.

---

1) $\displaystyle\int_a^b k \, dx = k(b - a)$ (k; a constant function)

2) $\displaystyle\int_a^b kf(x)dx = k\int_a^b f(x)dx$

3) $\displaystyle\int_a^b (f(x) + g(x))dx = \int_a^b f(x)dx + \int_a^b g(x)dx$

4) $\displaystyle\int_a^b (f(x) - g(x))dx = \int_a^b f(x)dx - \int_a^b g(x)dx$

$\because$ 1)

$\displaystyle\int_a^b k \, dx = k(b - a)$

2) Let $\displaystyle\int f(x)dx = F(x) + C_1$

Then, $\displaystyle\int kf(x)dx = k\int f(x)dx = kF(x) + C$

$\displaystyle\therefore \int_a^b kf(x)\,dx = k[F(x)]_a^b = k\{F(b) - F(a)\} = k\int_a^b f(x)\,dx$

3), 4)   Let $\displaystyle\int f(x)dx = F(x) + C_1,\quad \int g(x)dx = G(x) + C_2$

Then, $\displaystyle\int (f(x) \pm g(x))dx = \int f(x)dx \pm \int g(x)dx = F(x) \pm G(x) + C$

$\displaystyle\therefore \int_a^b (f(x) \pm g(x))dx = [F(x) \pm G(x)]_a^b = \{F(b) \pm G(b)\} - \{F(a) \pm G(a)\}$

$$= \{F(b) - F(a)\} \pm \{G(b) - G(a)\}$$

$$= \int_a^b f(x)dx \pm \int_a^b g(x)dx$$

Note: (1) For a continuous function f on $[a, b]$ such that $f(x) \le 0$ $(a \le x \le b)$,

the area above the graph of f from a to b is $\displaystyle S = -\int_a^b f(x)\,dx$

If $\displaystyle\int f(x)dx = F(x) + C,$ then $S = -[F(x)]_a^b = -\big(F(b) - F(a)\big) = F(a) - F(b)$

(2) Consider the two functions f and g defined on $\lfloor a, b]$ with $f(x) \ge g(x)$.

The area of the region between the graphs is given by $\displaystyle\int_a^b (f - g)\,dx$.

Since the area under the graph of f is $\displaystyle\int_a^b f(x)\,dx$ and that under g is $\displaystyle\int_a^b g(x)\,dx$,

the area of the difference between the regions (shaded part) is $\displaystyle\int_a^b f(x)\,dx - \int_a^b g(x)dx$.

Therefore, the difference is $\displaystyle\int_a^b (f - g)\,dx = \int_a^b f(x)\,dx - \int_a^b g(x)dx$.

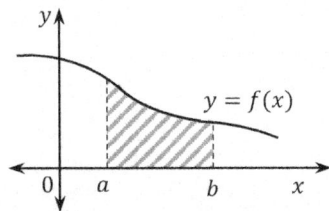
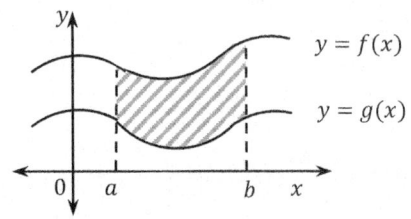

Example   Evaluate $\int_0^1 \{x^3 + (x^2 + 1)^4 x\}dx$.

$$\int_0^1 \{x^3 + (x^2 + 1)^4 x\}dx = \int_0^1 x^3 dx + \int_0^1 (x^2 + 1)^4 x dx$$

$$\int_0^1 x^3 dx = \left[\frac{1}{4}x^4\right]_0^1 = \frac{1}{4}(1 - 0) = \frac{1}{4}$$

$$\int (x^2 + 1)^4 x dx = \frac{1}{2}\int (x^2 + 1)^4 \cdot 2x dx = \frac{1}{2}\int t^4 dt = \frac{1}{2}\left(\frac{1}{5}t^5\right) + C = \frac{1}{10}t^5 + C$$

$$= \frac{1}{10}(x^2 + 1)^5 + C$$

$$\int_0^1 (x^2 + 1)^4 x dx = \left[\frac{1}{10}(x^2 + 1)^5\right]_0^1 = \frac{1}{10}(2^5 - 1) = \frac{31}{10}$$

Therefore, $\int_0^1 \{x^3 + (x^2 + 1)^4 x\}dx = \frac{1}{4} + \frac{31}{10} = \frac{10+124}{40} = \frac{134}{40}$

Example   Find the area of the region bounded by the curves $y = x^2 + 1$, $y = x^3$,
the vertical lines $x = 0$, and $x = 1$.

The area of the shaded region is

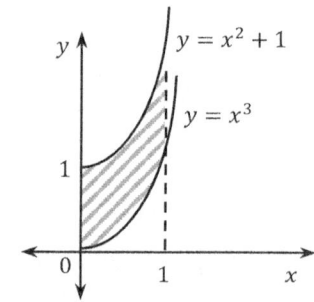

$$\int_0^1 (x^2 + 1)dx - \int_0^1 x^3 dx = \int_0^1 (x^2 + 1 - x^3)dx$$

$$= \left[\frac{1}{3}x^3 + x - \frac{1}{4}x^4\right]_0^1 = \frac{1}{3} + 1 - \frac{1}{4} = \frac{13}{12}$$

(2) Integral Additive Property

For a function $f(x)$ that is continuous on an interval containing the three points a, b, and c,

$$\int_a^b f(x)dx = \int_a^c f(x)dx + \int_c^b f(x)dx$$

∵ Let $F(x)$ be an antiderivative of $f(x)$.

Then, $\int f(x)\,dx = F(x) + C$

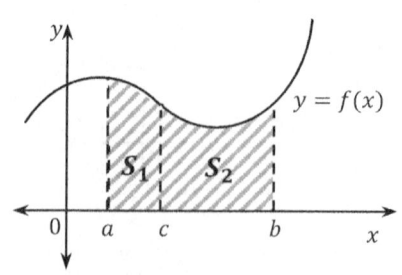

∴ $\int_a^c f(x)dx + \int_c^b f(x)dx = [F(x)]_a^c + [F(x)]_c^b$

$$= F(c) - F(a) + F(b) - F(c)$$

$$= F(b) - F(a)$$

$$= \int_a^b f(x)dx$$

We can break up one integral into the sum of two or more.

Example

(1) $\int_0^1 x^2 dx + \int_1^3 x^2 dx = \int_0^3 x^2 dx = \left[\frac{1}{3}x^3\right]_0^3 = \frac{1}{3}(3^3 - 0) = 9$

(2) $\int_0^2 x^2 dx = \int_0^1 x^2 dx + \int_1^2 x^2 dx = \left[\frac{1}{3}x^3\right]_0^1 + \left[\frac{1}{3}x^3\right]_1^2 = \frac{1}{3}(1^3 - 0) + \frac{1}{3}(2^3 - 1^3)$

$\qquad = \frac{1}{3} + \frac{7}{3} = \frac{8}{3}$

Note: $\int_0^2 x^2 dx = \int_0^3 x^2 dx + \int_3^2 x^2 dx = \int_0^3 x^2 dx - \int_2^3 x^2 dx = \left[\frac{1}{3}x^3\right]_0^3 - \left[\frac{1}{3}x^3\right]_2^3$

$\qquad = \frac{1}{3}(3^3 - 0) - \frac{1}{3}(3^3 - 2^3) = \frac{8}{3}$

Therefore, definite integrals do it no matter how the three points a, b, and c are arranged.

(3) Comparison Properties

1) Basic Inequality for Integrals

> If f and g are continuous on $[a, b]$ and if $f(x) \leq g(x)$ for all x in $[a, b]$, then
>
> $$\int_a^b f(x)dx \leq \int_a^b g(x)dx$$

$\because$ Let $a = x_0 < x_1 < x_2 < x_3 < \cdots\cdots < x_n = b$ be an arbitrary partition of $[a, b]$ and

$\Delta x = \frac{b-a}{n}$ be the common interval.

Since $f(x_i) \leq g(x_i)$, $f(x_i)\Delta x \leq g(x_i)\Delta x$

$\therefore \sum_{i=1}^n f(x_i)\Delta x \leq \sum_{i=1}^n g(x_i)\Delta x \qquad \therefore \lim_{n\to\infty} \sum_{i=1}^n f(x_i)\Delta x \leq \lim_{n\to\infty} \sum_{i=1}^n g(x_i)\Delta x$

Therefore, $\qquad \int_a^b f(x)dx \leq \int_a^b g(x)dx$

Another Approach:

$\int_a^b g(x)dx - \int_a^b f(x)dx = \int_a^b (g(x) - f(x))dx$

Let $S_t = \int_a^t (g(x) - f(x))dx$

Then, $S_t' = g(t) - f(t)$

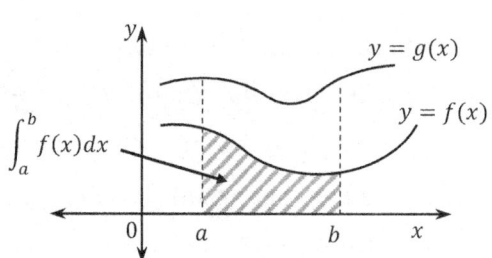

Since $g(x) \geq f(x)$ for all x in $[a, b]$, $S_t' \geq 0$

$\therefore$ S_t is an increasing function, and $S_b \geq S_a$

$\therefore$ $\displaystyle\int_a^b g(x)dx - \int_a^b f(x)dx = \int_a^b (g(x) - f(x))dx = S_b - S_a \geq 0$

Therefore, $\displaystyle\int_a^b g(x)dx \geq \int_a^b f(x)dx$

2) Boundedness Property

> If f is integrable on $[a, b]$ and if $m \leq f(x) \leq M$ for all x in $[a, b]$, then
>
> $$m(b - a) \leq \int_a^b f(x)dx \leq M(b - a)$$

From the Figure, we see $m(b - a)$ is the area of the small rectangle,

$M(b - a)$ is the area of the large rectangle, and

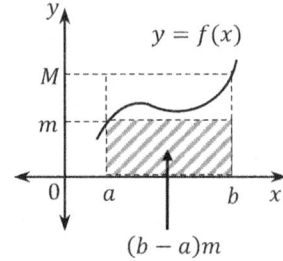

$\displaystyle\int_a^b f(x)dx$ is the area under the curve.

$\because$ Since $f(x) \leq M$, $\displaystyle\int_a^b f(x)dx \leq \int_a^b M dx$

Since $\displaystyle\int_a^b M dx = [Mx]_a^b = M(b - a)$, $\displaystyle\int_a^b f(x)dx \leq M(b - a)$

Since $m \leq f(x)$, $\displaystyle\int_a^b m dx \leq \int_a^b f(x)dx$

Since $\displaystyle\int_a^b m dx = [mx]_a^b = m(b - a)$, $m(b - a) \leq \displaystyle\int_a^b f(x)dx$

Therefore, $m(b - a) \leq \displaystyle\int_a^b f(x)dx \leq M(b - a)$

2. Functions defined by Integrals

(1) Differentiating a Definite Integral

Fundamental Theorem of Calculus

Let f be integrable on $[a, b]$ and x be any point in $[a, b]$.

Then, we can define the function G by the equation $G(x) = \displaystyle\int_a^x f(t)dt$

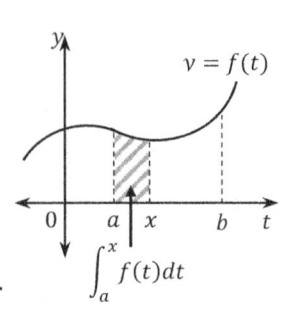

We use t as the variable for integration and x as a limit of integration.

Consider the derivative $G'(x)$.

By the fundamental theorem, $G' = f$ and so $G'(x) = f(x)$ for x in $[a, b]$.

This gives us the following formula:

$$\frac{d}{dx}\left[\int_a^x f(t)dt\right] = f(x)$$

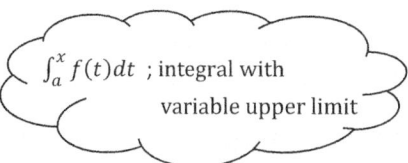

$\int_a^x f(t)dt$; integral with variable upper limit

That is, the derivative of the definite integral of a continuous function with respect to its upper limit of integration is equal to the value of the integrand function at the upper limit.

∵ When $G(x) = \int_a^x f(t)dt$, show that $G'(t) = \lim\limits_{h \to 0} \dfrac{G(x+h) - G(x)}{h} = f(x)$

$$G(x+h) - G(x) = \int_a^{x+h} f(t)dt - \int_a^x f(t)dt = \int_a^{x+h} f(t)dt + \int_x^a f(t)dt = \int_x^{x+h} f(t)dt$$

For $h > 0$, let m and M be the minimum and maximum values, respectively, of f on the interval $[x, x+h]$. Then, by Boundedness property,

$$mh \le \int_x^{x+h} f(t)dt \le Mh \quad ;\text{i.e.,} \quad mh \le G(x+h) - G(x) \le Mh$$

Since $h > 0$, $m \le \dfrac{G(x+h) - G(x)}{h} \le M$

Note that f is continuous, and m and M depends on h.

Thus, $m \to f(x)$ and $M \to f(x)$ as $h \to 0$

Therefore, $\lim\limits_{h \to 0} \dfrac{G(x+h) - G(x)}{h} = f(x)$

Similarly, for $h < 0$, we have the same result.

Generalized Fundamental Theorem of Calculus

Suppose f and h are continuous functions such that $G(x) = \int_a^x f(t)dt$ and h is defined.

If h is differentiable, then $G \circ h$ is also differentiable and we have $(G \circ h)'(x) = f(h(x))h'(x)$.

Since $(G \circ h)(x) = \int_a^{h(x)} f(t)dt$, $\dfrac{d}{dx}\left[\int_a^{h(x)} f(t)dt\right] = f(h(x))h'(x)$

Example

(1) $\dfrac{d}{dx}\left[\int_1^x t^2 dt\right]$

By the Fundamental Theorem, $\dfrac{d}{dx}\left[\int_1^x t^2 dt\right] = x^2$

Another Approach:

$$\int_1^x t^2\,dt = \left[\frac{1}{3}t^3\right]_1^x = \frac{1}{3}(x^3 - 1^3) = \frac{1}{3}x^3 - \frac{1}{3}$$

$$\therefore \frac{d}{dx}\left[\int_1^x t^2\,dt\right] = \frac{d}{dx}\left(\frac{1}{3}x^3 - \frac{1}{3}\right) = \frac{1}{3}\cdot 3x^2 = x^2$$

(2) $\dfrac{d}{dx}\left[\displaystyle\int_1^x \dfrac{t^{\frac{3}{2}}}{\sqrt{t^2+3}}\,dt\right]$

By the Fundamental Theorem, $\qquad \dfrac{d}{dx}\left[\displaystyle\int_1^x \dfrac{t^{\frac{3}{2}}}{\sqrt{t^2+3}}\,dt\right] = \dfrac{x^{\frac{3}{2}}}{\sqrt{x^2+3}}$

Note: $\displaystyle\int_x^{x+a} f(t)\,dt = \int_0^{x+a} f(t)\,dt - \int_0^x f(t)\,dt$

$$\therefore \frac{d}{dx}\left[\int_x^{x+a} f(t)\,dt\right] = \frac{d}{dx}\left[\int_0^{x+a} f(t)\,dt\right] - \frac{d}{dx}\left[\int_0^x f(t)\,dt\right] = f(x+a) - f(x)$$

That is, $\qquad \boxed{\dfrac{d}{dx}\left[\displaystyle\int_x^{x+a} f(t)\,dt\right] = f(x+a) - f(x) \quad (a;\text{constant})}$

Note: $\qquad \boxed{\displaystyle\lim_{x\to a}\frac{1}{x-a}\int_a^x f(t)\,dt = f(a) \; ; \quad \lim_{x\to 0}\frac{1}{x}\int_a^{x+a} f(t)\,dt = f(a)}$

$\because$ For a continuous function $f(x)$, let $\displaystyle\int f(x)\,dx = F(x)$.

$$\lim_{x\to a}\frac{1}{x-a}\int_a^x f(t)\,dt = \lim_{x\to a}\frac{1}{x-a}\{F(x) - F(a)\} = \lim_{x\to a}\frac{F(x) - F(a)}{x-a} = F'(a) = f(a)$$

$$\lim_{x\to 0}\frac{1}{x}\int_a^{x+a} f(t)\,dt = \lim_{x\to a}\frac{1}{x}\{F(x+a) - F(a)\} = \lim_{x\to a}\frac{F(x+a) - F(a)}{(x+a) - a} = F'(a) = f(a)$$

Example

(1) $\dfrac{d}{dx}\left[\displaystyle\int_x^{x+1}(3t^2 + t)\,dt\right] = 3(x+1)^2 + (x+1) - (3x^2 + x) = 6x + 4$

(2) $\dfrac{d}{dx}\left[\displaystyle\int_x^{x+1}\log t\,dt\right] \; (x>0) = \log(x+1) - \log x = \log\dfrac{x+1}{x}$

(2) Mean Value Theorem for Integrals

> If f is continuous function on a closed interval $[a, b]$, then there is a number c between a and b such that $\displaystyle\int_a^b f(t)dt = f(c)(b - c)$

$\because$ Let $\displaystyle\int_a^x f(t)dt = G(x), \qquad a \le x \le b$

By the Mean Value Theorem for derivatives, there is a point c in (a, b) such that

$G'(c) = \dfrac{G(b) - G(a)}{b - a}$; i.e., $G(b) - G(a) = G'(c)(b - a)$

Since $G(b) - G(a) = \displaystyle\int_a^b f(t)dt - \int_a^a f(t)dt = \int_a^b f(t)dt, \quad \int_a^b f(t)dt = G'(c)(b - a)$

Note:
$$\int_a^b f(t)dt = f(c)(b - a) \quad \Rightarrow \quad f(c) = \frac{\int_a^b f(t)dt}{b - a}$$

The number $\dfrac{\int_a^b f(t)dt}{b - a}$ is called the *mean value* or *average value* of f on $[a, b]$.

6-4 Evaluating Definite Integrals

1. Substitution Rule

$$\int f(x)dx = F(x) + C$$
$$\Rightarrow \int_a^b f(x)dx = [F(x)]_a^b = F(b) - F(a)$$

To evaluate definite integrals,

first, find an indefinite integral and then apply the Fundamental Theorem.

For example, (1) calculate $\displaystyle\int_0^2 (3x + 2)^2\, dx$.

First, let $3x + 2 = t$

Then $3dx = dt$; $dx = \dfrac{1}{3}dt$

$\therefore \displaystyle\int (3x+2)^2 dx = \int t^2 \cdot \frac{1}{3}dt = \frac{1}{3}\int t^2 dt = \frac{1}{3}\cdot\frac{1}{3}t^3 + C = \frac{1}{9}(3x+2)^3 + C$

By the Fundamental Theorem,

$\displaystyle\int_0^2 (3x+2)^2\, dx = \left[\frac{1}{9}(3x+2)^3\right]_0^2 = \frac{1}{9}(8^3 - 2^3) = \frac{504}{9} = 56$

(2) Calculate $\displaystyle\int \cos(2x + 1)\,dx$.

First, let $2x + 1 = t$

Then $2dx = dt$; $dx = \dfrac{1}{2}dt$

$$\therefore \int \cos(2x + 1)\,dx = \int \cos t \cdot \frac{1}{2}dt = \frac{1}{2}\int \cos t\,dt = \frac{1}{2}\sin t + C = \frac{1}{2}\sin(2x + 1) + C$$

Note: Chain Rule for derivative is applied for integrals.

Power Rule: $\displaystyle\int x^n dx = \frac{1}{n+1}x^{n+1} + C$

$$\int \big(f(x)\big)^n f'(x)dx = \int t^n dt = \frac{1}{n+1}t^{n+1} + C = \frac{1}{n+1}\big(f(x)\big)^{n+1} + C$$

---

(1) Substitution Rule for Indefinite Integrals

Let $\displaystyle\int f(x)dx = F(x) + C$ (F is an antiderivative of f) and g be a differentiable function.

If $g(x) = t$, then

$$\int f\big(g(x)\big)g'(x)dx = \int f(t)dt = F(t) + C = F\big(g(x)\big) + C$$

Note: $\dfrac{d}{dx}\big[F(g(x)) + C\big] = F'(g(x))g'(x) = f(g(x))g'(x)$

$$\therefore\ F\big(g(x)\big) + C = \int f\big(g(x)\big)g'(x)dx$$

Example   Evaluate $\displaystyle\int_0^{\sqrt{\pi}} x\sin^4(x^2)\cos(x^2)\,dx$.

Let $\sin(x^2) = t$

Then, $\cos(x^2) \cdot 2x\,dx = dt$; $\cos(x^2) \cdot x\,dx = \dfrac{1}{2}dt$

$$\int x\sin^4(x^2)\cos(x^2)\,dx = \int t^4 \cdot \frac{1}{2}dt = \frac{1}{2}\int t^4 dt = \frac{1}{2}\cdot\frac{1}{5}t^5 + C = \frac{1}{10}\sin^5(x^2) + C$$

$$\therefore \int_0^{\sqrt{\pi}} x\sin^4(x^2)\cos(x^2)\,dx = \frac{1}{10}[\sin^5(x^2)]_0^{\sqrt{\pi}} = \frac{1}{10}(\sin^5\pi - \sin^5 0) = 0$$

(The limits 0 and $\sqrt{\pi}$ apply to x, not t.)

(2) Substitution Rule for Definite Integrals

> Let g have a continuous derivative on $[a, b]$ and let f be continuous on the range of g.
>
> Then, $\displaystyle\int_a^b f(g(x))\, g'(x)dx = \int_{g(a)}^{g(b)} f(t)dt$

That is, when $g(x) = t$ is monotone increasing or decreasing, and g is a differentiable function,

and if $g(a) = m$, and $g(b) = n$, then $\displaystyle\int_a^b f(g(x))\, g'(x)dx = \int_m^n f(t)dt$

$\because$ Let $\displaystyle\int f(x)dx = F(x) + C$

Then, $\displaystyle\int_{g(a)}^{g(b)} f(t)dt = [F(t)]_{g(a)}^{g(b)} = F(g(b)) - F(g(a))$ by the Fundamental Theorem.

By the substitution theorem for indefinite integrals,

$$\int f(g(x))g'(x)dx = F(g(x)) + C$$

Again, by the Fundamental Theorem,

$$\int_a^b f(g(x))\, g'(x)dx = \left[F(g(x))\right]_a^b = F(g(b)) - F(g(a))$$

Example Evaluate.

(1) $\displaystyle\int_2^4 \sqrt{x-2}\, dx$

Let $x - 2 = t$ Then, $dx = dt$

$t = 0$ when $x = 2$

$t = 2$ when $x = 4$

$\therefore \displaystyle\int_2^4 \sqrt{x-2}\, dx = \int_0^2 \sqrt{t}\, dt = \int_0^2 t^{\frac{1}{2}}\, dt = \left[\frac{2}{3}t^{\frac{3}{2}}\right]_0^2 = \frac{2}{3}2^{\frac{3}{2}} = \frac{2}{3}2\sqrt{2} = \frac{4}{3}\sqrt{2}$

(2) $\displaystyle\int_0^3 2xe^{x^2}\, dx$

Let $x^2 = t$ Then, $2xdx = dt$

$t = 0$ when $x = 0$

$t = 9$ when $x = 3$

$\therefore \displaystyle\int_0^3 2xe^{x^2}\, dx = \int_0^9 e^t\, dt = [e^t]_0^9 = e^9 - e^0 = e^9 - 1$

2. Symmetry Rule

If $f(-x) = f(x)$ for any x, then f is an even function (the graph of f is symmetric with respect to the y-axis), whereas if $f(-x) = -f(x)$ for any x, then f is an odd function (the graph of f is symmetric with respect to the origin).

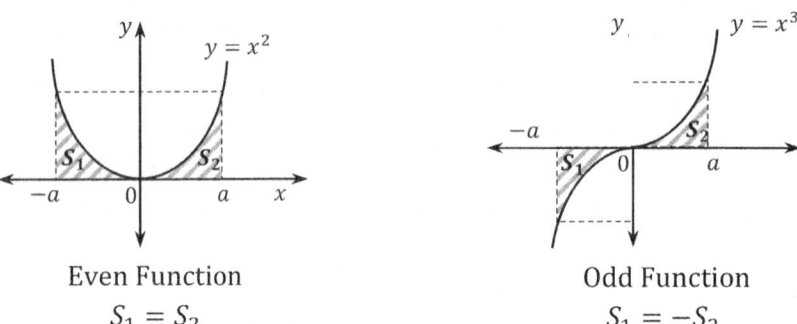

Even Function	Odd Function
$S_1 = S_2$	$S_1 = -S_2$

<u>Symmetry Theorem</u>

(1) If f is an even function, then $\displaystyle\int_{-a}^{a} f(x)\,dx = 2\int_{0}^{a} f(x)\,dx$

(2) If f is an odd function, then $\displaystyle\int_{-a}^{a} f(x)\,dx = 0$

$\because \displaystyle\int_{-a}^{a} f(x)\,dx = \int_{-a}^{0} f(x)\,dx + \int_{0}^{a} f(x)\,dx$

For $\displaystyle\int_{-a}^{0} f(x)\,dx,$ let $x = -t\,;\ dx = -dt$

$\therefore \displaystyle\int_{-a}^{0} f(x)\,dx = \int_{a}^{0} f(-t)\,(-dt) = -\int_{a}^{0} f(-t)\,dt = \int_{0}^{a} f(-t)\,dt$

(1) If $f(x)$ is an even function, then $f(-t) = f(t)$.

$\therefore \displaystyle\int_{-a}^{0} f(x)\,dx = \int_{0}^{a} f(-t)\,dt = \int_{0}^{a} f(t)\,dt$

$\therefore \displaystyle\int_{-a}^{a} f(x)\,dx = \int_{-a}^{0} f(t)\,dt + \int_{0}^{a} f(x)\,dx = \int_{0}^{a} f(x)\,dx + \int_{0}^{a} f(x)\,dx = 2\int_{0}^{a} f(x)\,dx$

(2) If $f(x)$ is an odd function, then $f(-t) = -f(t)$.

$\therefore \displaystyle\int_{-a}^{0} f(x)\,dx = \int_{0}^{a} f(-t)\,dt = -\int_{0}^{a} f(t)\,dt$

$\therefore \displaystyle\int_{-a}^{a} f(x)\,dx = -\int_{0}^{a} f(t)\,dt + \int_{0}^{a} f(x)\,dx = -\int_{0}^{a} f(x)\,dx + \int_{0}^{a} f(x)\,dx = 0$

Example

$$\int_{-\pi}^{\pi} \sin x \, dx = 0 \quad (\because \ \sin x \text{ is an odd function})$$

$$\int_{-\frac{\pi}{2}}^{\frac{\pi}{2}} \cos x \, dx = 2 \int_{0}^{\frac{\pi}{2}} \cos x \, dx \quad (\because \ \cos x \text{ is an even function})$$

$$= [2 \sin x]_{0}^{\frac{\pi}{2}} = 2 \left(\sin \frac{\pi}{2} - \sin 0 \right) = 2(1 - 0) = 2$$

$$\int_{-1}^{1} (x^5 + 4x^3 + 2x^2 + 1) \, dx = \int_{-1}^{1} (x^5 + 4x^3) \, dx + \int_{-1}^{1} (2x^2 + 1) \, dx$$

$$= 0 + 2 \int_{0}^{1} (2x^2 + 1) \, dx$$

$$(\because \ x^5 \text{ and } x^3 \text{ are odd functions, and } x^2 \text{ and constant are even functions})$$

$$= 2 \left[\frac{2}{3} x^3 + x \right]_{0}^{1} = 2 \left(\frac{2}{3} + 1 \right) = \frac{10}{3}$$

3. Periodicity Rule

A function f is *periodic* if there is a number p such that $f(x + p) = f(x)$ for all real numbers x in the domain of f.

$$\boxed{\text{If } f \text{ is a periodic with period } p, \ \text{ then } \int_{a+p}^{b+p} f(x) \, dx = \int_{a}^{b} f(x) \, dx}$$

$\because$ Let $x - p = t \quad$ Then, $x = t + p \ ; \ dx = dt$

$$\int_{a+p}^{b+p} f(x) \, dx = \int_{a}^{b} f(t + p) \, dt = \int_{a}^{b} f(t) \, dt \quad (\because f(x) \text{ is periodic with period } p)$$

$$= \int_{a}^{b} f(x) \, dx$$

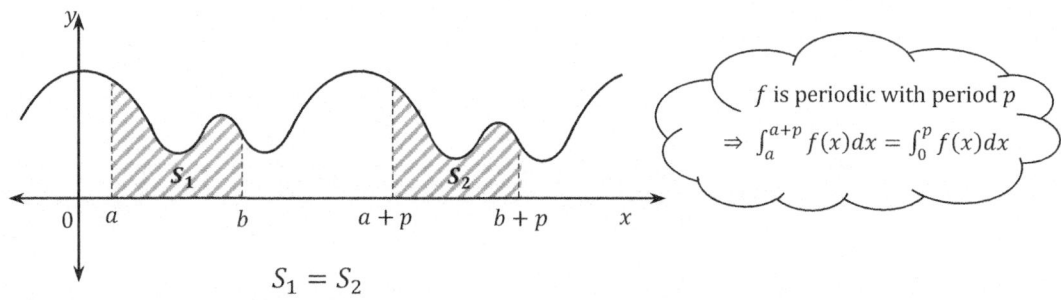

$S_1 = S_2$

Example

(1) $\displaystyle\int_0^{2\pi} |\sin x|\,dx = \int_0^{\pi} |\sin x|\,dx + \int_{\pi}^{2\pi} |\sin x|\,dx$

$\displaystyle\qquad\qquad = \int_0^{\pi} |\sin x|\,dx + \int_0^{\pi} |\sin x|\,dx = 2\int_0^{\pi} |\sin x|\,dx$

$\displaystyle\qquad\qquad = 2\int_0^{\pi} \sin x\,dx = 2[-\cos x]_0^{\pi} = -2(\cos\pi - \cos 0) = -2(-1-1) = 4$

(2) $\displaystyle\int_1^{1+\pi} |\sin x|\,dx = \int_0^{\pi} |\sin x|\,dx = \int_0^{\pi} \sin x\,dx = [-\cos x]_0^{\pi} = -(\cos\pi - \cos 0) = 2$

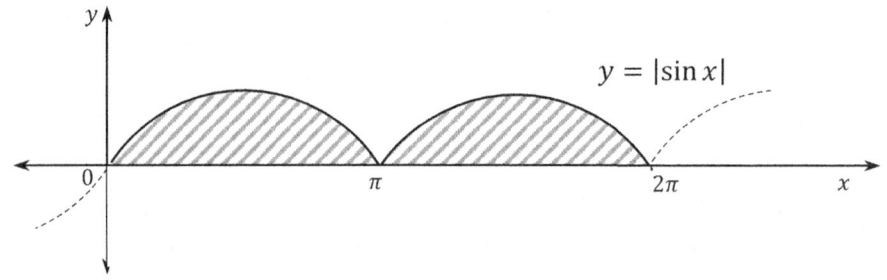

Exercises

#1 Use the fundamental theorem of calculus to evaluate each definite integral.

(1) $\int_1^3 3x^2 \, dx$

(2) $\int_{-1}^2 (2x - 6x^2) \, dx$

(3) $\int_0^1 \sqrt{x} \, dx$

(4) $\int_0^\pi \cos x \, dx$

(5) $\int_0^{\frac{\pi}{2}} \sin 2x \, dx$

(6) $\int_0^2 e^{-x} \, dx$

(7) $\int_{-1}^1 (x - 1)(x^2 + x + 1) \, dx$

(8) $\int_0^2 (y + 1)(y^2 - 1) \, dy$

(9) $\int_1^2 \frac{x^2 + 2}{x} \, dx$

(10) $\int_1^2 \frac{1}{x^2 + x} \, dx$

(11) $\int_1^2 \frac{x}{x^2 + 1} \, dx$

(12) $\int_0^{\frac{\pi}{3}} \tan^2 x \, dx$

(13) $\int_0^{\frac{\pi}{2}} (e^{4x} - \sin^2 x) \, dx$

(14) $\int_0^1 x e^{-x^2} \, dx$

(15) $\int_0^2 (2x + 1)\sqrt{x^2 + x} \, dx$

(16) $\int_1^0 5 \, dx$

(17) $\int_4^1 \left(\sqrt{x} + \frac{1}{\sqrt{x}} + 1 \right) dx$

(18) $\int_0^1 \left(\frac{1}{\sqrt{x+1} - \sqrt{x}} \right) dx$

(19) $\int_{-1}^{-2} \left(\frac{2}{x^2} + \frac{1}{x} \right) dx$

(20) $\int_0^1 \frac{(x+2)^2}{x+1} \, dx$

(21) $\int_2^4 \frac{1}{x^2 - 3x + 2} \, dx$

(22) $\int_1^3 \frac{x^2(x^2 + 2x + 4)}{x + 2} \, dx + \int_3^1 \frac{4(y^2 + 2y + 4)}{y + 2} \, dy$

(23) $\int_{-\pi}^{\pi} (\sin x + \cos x)^2 \, dx$

(24) $\int_0^{\frac{\pi}{4}} (\sin^3 2x \cos 2x) \, dx$

(25) $\int_0^{2\pi} \cos 3x \cos 2x \, dx$

(26) $\frac{1}{\pi} \int_{-\pi}^{\pi} \sin 3x \, (\sin 3x + \sin 5x) \, dx$

(27) $\int_0^1 \frac{x^2}{x+1} \, dx - \int_0^1 \frac{1}{x+1} \, dx$

(28) $\int_0^{\log 2} \frac{e^{3x}}{e^x + 1} \, dx - \int_{\log 2}^0 \frac{1}{e^t + 1} \, dt$

#2 Evaluate.

(1) $\displaystyle\lim_{n\to\infty} \frac{1}{n\sqrt{n}} \sum_{k=1}^n (\sqrt{n} + \sqrt{k})$

(2) $\displaystyle\lim_{n\to\infty} \sum_{i=1}^n \frac{3}{n} \sin^2 \frac{\pi i}{n}$

(3) $\displaystyle\lim_{n\to\infty} \sum_{j=1}^n \frac{j}{n^2} \cos \frac{\pi j^2}{3n^2}$

(4) $\displaystyle\lim_{n\to\infty} \frac{1}{n^2} \sum_{l=1}^n l e^{\frac{l}{n}}$

(5) $\displaystyle\lim_{n\to\infty} \sum_{i=1}^{2n} \frac{1}{2n + i}$

(6) $\displaystyle\lim_{n\to\infty} \frac{1}{n^{k+1}} \sum_{i=1}^n i^k$

(7) $\displaystyle\lim_{n\to\infty} \log \left\{ \frac{(2n)!}{n^n n!} \right\}^{\frac{1}{n}}$

(8) $\displaystyle\lim_{n\to\infty} \int_0^1 \left(\sum_{i=1}^n \frac{1}{i} x^i \right) dx$

(9) $\lim\limits_{n\to\infty}\dfrac{1}{n}\left\{\log\left(1+\dfrac{1}{n}\right)+\log\left(1+\dfrac{2}{n}\right)+\cdots\cdots+\log\left(1+\dfrac{n}{n}\right)\right\}$

(10) $\lim\limits_{n\to\infty}\dfrac{\pi}{n^2}\left(\sin\dfrac{\pi}{n}+2\sin\dfrac{2\pi}{n}+3\sin\dfrac{3\pi}{n}+\cdots\cdots+n\sin\dfrac{n\pi}{n}\right)$

(11) $\lim\limits_{n\to\infty}\dfrac{1}{n^3}\left\{\sqrt{n^2-1^2}+2\sqrt{n^2-2^2}+\cdots\cdots+(n-1)\sqrt{n^2-(n-1)^2}\right\}$

(12) $\lim\limits_{n\to\infty}\left(\dfrac{n}{1^2+3n^2}+\dfrac{n}{2^2+3n^2}+\dfrac{n}{3^2+3n^2}+\cdots\cdots+\dfrac{n}{n^2+3n^2}\right)$

(13) $\lim\limits_{n\to\infty}\dfrac{1}{n}\left\{\left(\dfrac{n}{n}\right)^3+\left(\dfrac{n+1}{n}\right)^3+\left(\dfrac{n+2}{n}\right)^3+\cdots\cdots+\left(\dfrac{2n-1}{n}\right)^3\right\}$

#3 Suppose $\int_a^b \sin x\,dx = A$, $\int_b^c \sin x\,dx = B$, and $\int_a^c \sin\left(x+\dfrac{\pi}{4}\right)dx = C$.

Use properties of definite integrals to calculate each of the following integrals:

(1) $\int_a^a \sin x\,dx$

(2) $\int_b^a \sin x\,dx$

(3) $\int_a^c \sin x\,dx$

(4) $\int_a^c \left\{\sin x + 2\sin\left(x+\dfrac{\pi}{4}\right)\right\}dx$

(5) $\int_a^c \cos x\,dx$

#4 Find the value.

(1) When $f(a) = \int_a^{a+1}(x^2+a^2)dx$, find the value of $\int_0^1 f(x)dx$.

(2) For a sequence $\{a_n\}$, $\displaystyle\sum_{i=1}^{n} a_i = \int_0^n (2x+3)dx$. Find the value of a_{10}.

(3) For a function $f(x) = ax^3 + bx + c$ such that i) $\lim\limits_{x\to 1}\dfrac{f(x)}{x-1} = 2$ and ii) $\int_0^1 f(x)dx = 2$,

Find the value of $a - b - c$. $(a, b, c;$ constants$)$

(4) When $f(a) = \int_0^a |x-1|dx$, find the value of $f'(2)$.

(5) When a function $f(x) = |x+1| + |x| + |x-1|$ has the minimum value m,

find the value of $\int_{-2}^m f(x)dx$.

(6) When $a_n = (\log 2)\int_0^n 2^x dx$ $(n = 0, 1, 2, \cdots\cdots)$, find the value of $\displaystyle\sum_{n=0}^{\infty}\dfrac{1}{1+a_n}$.

(7) When $f(x) = e^{-3x}$ and $g(x) = \dfrac{1}{1+x}$, find the value of $\int_0^{\log 2} g(f(x))dx$.

(8) For a function $f(x)$, $F(x) = \int f(x)dx$.

When $F(x) = xf(x) - 4x^2(x+1)$ and $f(1) = 10$, find the value of $\int_0^1 f(x)dx$.

#5 Show that $\int_m^n a(x-m)(x-n)dx = -\frac{a}{6}(n-m)^3$.

#6 Given $f(x) = |x-1|$, evaluate the following integrals:

(1) $\int_0^1 f(x)dx$

(2) $\int_0^3 f(x)dx$

(3) $\int_0^2 |x|f(x)dx$

#7 Suppose $f(x) = \begin{cases} x^2 + 2, & x \leq 1 \\ -x + 4, & x \geq 1 \end{cases}$.

Use the interval additive property to evaluate the integral.

(1) $\int_{-1}^1 f(x)dx$

(2) $\int_0^2 f(x)dx$

(3) $\int_0^5 f(x)dx$

#8 Find the limit.

(1) $\lim\limits_{x\to 1} \frac{1}{x-1}\int_1^x |t-2|dt$

(2) $\lim\limits_{x\to 3} \frac{1}{x-3}\int_3^x t^3 e^t dt$

(3) $\lim\limits_{x\to 1} \frac{1}{x-1}\int_1^x \sqrt{3+2t^3}dt$

(4) $\lim\limits_{x\to 2} \frac{1}{x-2}\int_2^x (t-1)(2t+3)dt$

(5) $\lim\limits_{x\to 2} \frac{1}{x^2-4}\int_2^x (3t^2+3t-2)dt$

(6) $\lim\limits_{x\to 0} \frac{1}{x}\int_0^x \frac{\cos t}{1+\sin t}dt$

(7) $\lim\limits_{h\to 0} \frac{1}{h}\int_1^{1+3h}(x\log x + x^2 e^x)dx$

(8) $\lim\limits_{h\to 0} \frac{1}{h}\int_{2-h}^{2+h} \frac{2x-1}{x^2+1}dx$

(9) $\lim\limits_{n\to\infty} \int_0^n (x^k e^{-x})dx$ $\left(\lim\limits_{x\to\infty} \frac{x^n}{e^x} = 0\right)$

#9 Find the indicated integrals.

(1) $\int_0^\pi |\sin 2x|dx$

(2) $\int_0^1 |e^x - 3|dx$

(3) $\int_{-3}^3 |(x-1)(x^2+x-1)|dx$

(4) $\int_{-3}^7 ||x-2|-3|dx$

(5) $\int_0^2 |x-a|dx$ $(a \geq 0)$

(6) $\int_0^2 |x - \sqrt{x}|dx$

(7) $\int_{-1}^4 \left|\frac{x-2}{x+2}\right|dx$

(8) $\int_0^\pi |\sin x + \cos x|dx$

(9) $\int_0^\pi |\sin x \cos x|dx$

(10) $\int_{-1}^1 |3^x - 2^x|dx$

#10 Find the derivative of a definite integral.

(1) $\frac{d}{dx}\left[\int_3^x (t^4 + 3t^2 - 5t + 2)dt\right]$

(2) $\frac{d}{dx}\left[\int_x^3 (\cot^2 t \sin t)dt\right]$, $\frac{\pi}{2} < x < \frac{3\pi}{2}$

(3) $\frac{d}{dx}\left[\int_0^{x^2} (2t-1)dt\right]$

(4) $\frac{d}{dx}\left[\int_{3x}^2 \sqrt{t^2+3}dt\right]$

#11 Find the function $f(x)$.

(1) Find $f(x)$ such that $f(x) = 3x + \int_0^3 f(x)dx$.

(2) Find $f(x)$ such that $f(x) = \sin x + 2\int_0^{\frac{\pi}{2}} f(x)\cos x\, dx$.

(3) The slope of a tangent line at a point (x, y) on the graph of $y = f(x)$ is $e^x + a$.

    Find $f(x)$ such that $f(0) = 1$ and $\int_0^1 f(x)dx = e + 1$.

(4) Find $f(x)$ such that i) $\int_0^x e^t\{f(t) + f'(t)\}dt = f(x)f'(x) - 3$,

                ii) $f(x) > 0$, and

                iii) $f(0) = 3$.

(5) Find $f(x)$ such that $f(x) = x + 1 + \int_0^2 g(t)dt$ where $g(x) = 3x - 2 + \int_0^1 f(t)dt$.

(6) Find $f(x)$ such that $f(x) = 1 + \int_{-1}^1 (x^2 - t)f(t)dt$.

(7) For a continuous function $f(x)$ such that $f(x) = \int_0^x f(t)dt + 1$, find $f(x)$. $(f(x) \neq 0)$

(8) Find $f(x)$ such that $f(x) = \tan x - x - \int_\pi^x f'(t)\tan^2 t\, dt$.

(9) Find $f(x)$ and $g(x)$, and the constants a and b such that

    i) $\int_1^x \{f(t) - g(t)\}dt = 3x^2 - 2x + a$ and

    ii) $\int_1^x \{f(t) + g(t)\}dt = 2x^3 - x^2 + 2x + b$.

(10) Find a function $f(x)$ and a constant a such that $\int_a^{\log x} f(u)du = x^2 - 2x$. $(x > 0)$

(11) When a differentiable function $f(x)$ satisfies $\int_0^x (x - t)f(t)dt = e^x + ax + b$ for any real

    number x, find $f(x)$ and the constants a and b.

(12) When a differentiable function $f(x)$ satisfies $\int_a^x f(t)dt = e^{2x} - 4e^x + 3$ for any real

    number x, find $f(x)$ and the constant a.

(13) When a differentiable function $f(x)$ satisfies $\int_{\frac{1}{2}}^x f(t)dt = \sin \pi x - a\cos 2\pi x + 1$ for any

    real number x, find $f(x)$ and the constant a.

#12 Solve the equation of x.

(1) $\sin\left\{\frac{\pi}{3}\log_x\left[\frac{d}{dx}[\int_2^x t^3 dt]\right]\right\} = x^2 - 3x$

(2) $\log_{x^3}\left\{\frac{d}{dx}[\int_x^{x+1} \frac{1}{2}(t^2 - t)dt]\right\} = x^2 - 4x + \frac{10}{3}$

#13 Find the value.

(1) When $f(x)$ and $f'(x)$ are continuous functions such that

$f(x) = x^2 e^x - x + \int_0^x (x-t)f'(t)dt$, find the value of $f'(1) - f(1)$.

(2) When $f(x) = \int_0^x t \sin t \, dt \quad (0 \le x \le 2\pi)$,

1) Find the value of x so that $f(x)$ is a local extreme value.

2) Find the limit: $\displaystyle\lim_{h \to 0} \frac{f(\pi+2h)-f(\pi)}{h}$.

(3) Find the value of a so that $f(a) = \int_0^1 (e^x + ax)^2 dx$ has the minimum value.

#14 Use the method of substitution in definite integrals to evaluate each of the following:

(1) $\int_1^e x^2 e^x dx$

(2) $\int_0^\pi x \sin x \, dx$

(3) $\int_0^2 x^2 \sqrt{x^3 + 1} dx$

(4) $\int_1^3 \frac{2x}{1+x^2} dx$

(5) $\int_{\frac{\pi^2}{9}}^{\frac{\pi^2}{4}} \frac{\sin\sqrt{x}}{\sqrt{x}} dx$

(6) $\int_0^{\frac{\pi}{2}} \cos^3 x \sin x \, dx$

(7) $\int_0^{\frac{\pi}{2}} \frac{\sin^3 x}{1+\cos x} dx$

(8) $\int_0^{\sqrt{\pi}} x^3 \cos x^2 \, dx$

(9) $\int_0^\pi x|\cos x|dx$

(10) $\int_{\frac{\pi}{6}}^{\frac{\pi}{2}} \cos x \log\left(\frac{1}{\sin^2 x}\right) dx$

(11) $\int_1^e \frac{(\log x)^3}{x} dx$

(12) $\int_1^4 \frac{1}{\sqrt{x}} e^{\sqrt{x}} dx$

(13) $\int_1^e \frac{\log x}{x(\log x+1)^2} dx$

(14) $\int_0^1 \frac{x\log(1+x^2)}{1+x^2} dx$

(15) $\int_0^1 \frac{1}{e^x+1} dx$

(16) $\int_{\log 2}^1 \frac{1}{e^x-e^{-x}} dx$

(17) $\int_0^a \sqrt{a^2 - x^2} dx \quad (a > 0)$

(18) $\int_0^a \frac{1}{x^2+a^2} dx \quad (a \ne 0)$

#15 Find the value.

(1) For a differentiable function $f(x)$ such that $\int_0^1 f(x)dx = 6$, find the value of $\int_0^1 x^2 f(x^3)dx$.

(2) For a continuous function $f(x)$ such that $f(x) + f(-x) = x^2 - 1$,

find the value of $\int_{-1}^1 f(x)dx$.

(3) For a function $f(x) = 2x \log x + x^2 - 5x + 4$,

find the value of $\int_2^5 f(x)dx - \int_4^5 f(x)dx + \int_1^2 f(x)dx$.

(4) Let S_a be the area surrounded by the tangent line at $x = a$ on the graph of $y = e^x$, x-axis,

and a vertical line $x = a$. Find the value of $\displaystyle\lim_{n \to \infty} \sum_{i=1}^n \frac{1}{n} S_{\frac{n+3i}{n}}$.

(5) For a differentiable function $f(x)$ such that $f(1) = -1$ and $f'(1) = 2$,

find the value of $\lim\limits_{x \to 1} \frac{1}{x-1} \int_1^x \{f(t)\}^2 f'(t) dt$.

(6) For a differentiable function $f(x)$ on an open interval $(1, 3)$, $f(x)$ and $f'(x)$ are continuous on the closed interval $[1, 3]$. When $f(1) = 0$, and the minimum and maximum values of $f'(x)$ in the interval $[1, 3]$ are 2 and 5, respectively, find the range of $\int_1^3 f(x) dx$.

(7) The graph of a cubic function $y = f(x)$ is shown as the Figure. When $f(x)$ has a local maximum value 2 at $x = 1$, a local minimum value -2 at $x = 4$, and $f(0) = -2$, find the value of $\int_0^4 |f'(x)| dx$.

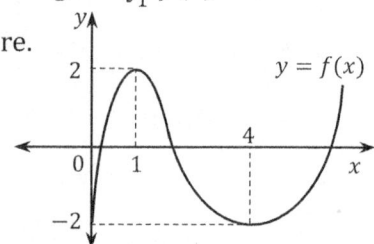

(8) For a positive number a, a function $f(x) = -2x(x + a)(x - a)$ has a local maximum value at $x = b$. When $A = \int_{-b}^a f(x) dx$ and $B = \int_b^{a+b} f(x - b) dx$, find the value of $\int_{-b}^a |f(x)| dx$.

#16 Use symmetry to help you evaluate the given integral.

(1) $\int_{-2}^2 (x^5 - 3x^2 + 4x) dx$

(2) $\int_{-2}^0 (x^3 + 2x) dx - \int_2^0 (x^3 + 2x) dx$

(3) $\int_{-\pi}^\pi (\sin x + \cos x) dx$

(4) $\int_{-\pi}^\pi \cos\left(\frac{x}{2}\right) dx$

(5) $\int_{-3}^3 \frac{x^5}{x^2 + 2} dx$

(6) $\int_{-2}^2 (x \sin^2 x + x^3 - x^2) dx$

#17 Use periodicity to calculate $\int_0^{4\pi} |\cos x| dx$.

#18 Find the value.

(1) For a function $f(x) = ax^2 + bx + 1$ $(a \neq 0)$ such that $\int_{-\pi}^\pi f(x) \sin x \, dx = 0$, the graph of $y = f(x)$ and a line $y = x$ intersect only at one point. Find the value of $a + b$.

(2) For any real number x, a function $f(x)$ satisfies $f(x) = f(x + 2)$.

When $f(x) = |x|$ $(-1 \le x \le 1)$, find the value of $\int_0^2 e^{2x} f(x) dx$.

(3) For a function $f(x) = ax \log x + b$ such that

i) $\lim\limits_{x \to e} \frac{f(x) - f(e)}{x - e} = 4$ and

ii) $\int_1^e f(x) dx = \frac{1}{2} e(e + 1)$,

find the value of $a + b$.

(4) When $S_n = \sum\limits_{i=1}^n \int_i^{i+1} 3x^2 dx$, find the value of $\lim\limits_{n \to \infty} \frac{S_n}{n^3}$.

(5) Find the minimum value of the positive integer n such that $S_n = \int_0^{\frac{1}{2}} \left(\sum_{i=1}^{n} i x^{i-1} \right) dx \geq 0.99$

(6) For a sequence $\{a_n\}$, the nth term is $a_n = \int_n^{n+1} \left(-\frac{1}{2^n} \sin \pi x \right) dx$. Find the value of $\sum_{n=1}^{\infty} a_n$.

(7) When $S_n = \int_1^{e^n} \left(\frac{\log x}{x} \right) dx$ $(n = 1, 2, 3, \cdots\cdots)$, find the value of $\sum_{n=1}^{\infty} \frac{1}{\sqrt{S_n S_{n+1}}}$.

(8) For a positive integer n, $f(x) = \int_0^1 \frac{1}{n} x^n dx$. Find the value of $\sum_{n=1}^{100} f(n)$.

(9) When $f(x) = x^3 + x$, find the value of $\lim_{n \to \infty} \frac{1}{n} \sum_{i=1}^{n} f\left(1 + \frac{2i}{n}\right)$.

(10) Find a local extreme value of $f(x) = \int_1^x (1 - \log t) dt$. $(x > 0)$

(11) Let $F(x) = \int_0^x (1 - t)e^t dt$.

Find the maximum value of the function F defined on real numbers.

(12) For a function $(x) = x^3 + ax$, $\frac{d}{dx}\left(\int_0^x f(t)dt\right) = \int_1^x \left(\frac{d}{dx} f(x)\right) dx$.

Find the value of the constant a.

(13) When $f(x) = \int_0^x e^{t^2} dt$ and $g(x) = \log f'(x)$,

find the value of a such that $\int_0^2 g(x)dx = 2g(a)$.

(14) For a function $f(x)$ such that $\int_0^x f(t)dt = xe^{-x}$, find the value of $\lim_{h \to 0} \frac{f(1+2h)-f(1)}{h}$.

(15) For a function $f(x) = \int_x^{2x} e^{\frac{2t}{\pi}} \sin t \, dt$, find the value of $f'\left(\frac{\pi}{2}\right)$.

(16) Find the value of x such that $\frac{d}{dx}\int_0^x (x - t)\cos^2 t \, dt = \frac{1}{2}x + \frac{1}{4}$. $(0 < x < \pi)$

(17) The graph of the function $y = f(x)$

is shown as the Figure.

For a function $g(x) = \int_x^{x+2} f(x)dx$,

find the maximum value of $g(x)$.

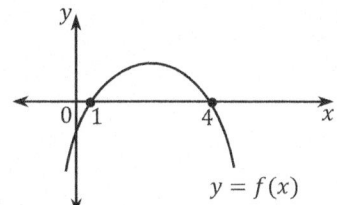

(18) For a function $f(x) = 3x^2 - x + 3\int_0^1 f(x)dx$, find the product of all solutions of the equation $f(x) = 0$.

(19) When $f(x) = x^3 - 2x + \int_0^2 g(t)dt$ and $g(x) = 3x^2 - 5 + \int_{-1}^1 f(t)dt$, find the value of $f(1) + g(-1)$.

(20) For a function $f(x)$ such that $\int_1^x f(t)dt = x^5 - 3x + a$,

find the value of $f'(-1) + a$. $(a; \text{constant})$

(21) Find the relationship between the real numbers a and b such that

$f(x) = \int_{x-1}^x (t^2 + at + b)dt > 0$ for any real number x.

(22) When a differentiable function $f(x)$ satisfies:

i) $f(-x) = -f(x)$ for any real number x and

ii) $f(2) = 3$,

find the value of $\int_{-2}^{2} f'(x)(3 - x)dx$.

(23) When a polynomial function $f(x)$ defined on real numbers satisfies:

i) $f(1) = 10$,

ii) $f(x) = \frac{1}{2}\int_{x}^{x+1} f(t)dt - \frac{1}{2}\int_{x}^{x-1} f(t)dt - \int_{0}^{1} f(t)dt$, and

iii) $f(x + y) + f(x - y) = 3\{f(x) + f(y)\}$,

find the value of $f'(1)$.

(24) For a function $f(x)$ such that $f(x + 2) = -f(x)$ and $\int_{0}^{2} f(x)dx = 10$,

find the value of $\int_{-2}^{4} f(x)dx$.

(25) For a polynomial function $f(x)$ such that $f'(-a) = f'(a)$ and $\int_{-a}^{a} f(x)dx = 4a$,

find the value of $f(0)$.

(26) For a polynomial function $g(x)$, let $f(x) = x^2 + ax + \int_{2}^{x} g(t)dt$.

When $f(x)$ is divided by $(x - 2)^2$, there is no remainder.

Find the remainder when $g(x)$ is divided by $x - 2$.

(27) For a quadratic function $f(x)$, $f(x) = \frac{12}{7}x^2 - 2x \int_{1}^{2} f(t)dt + \left[\int_{1}^{2} f(t)dt\right]^2$.

Find the value of $\int_{1}^{2} f(x)dx$.

Chapter 7. Applications of the Integral

7-1 The Area of a Plane Region

1. A Region between a Curve in the xy-Plane and x-axis

Let $y = f(x)$ be a continuous function on the interval $a \leq x \leq b$.

Consider the region S bounded by the curve of $y = f(x)$, $x = a$, $x = b$, and x-axis ($y = 0$).

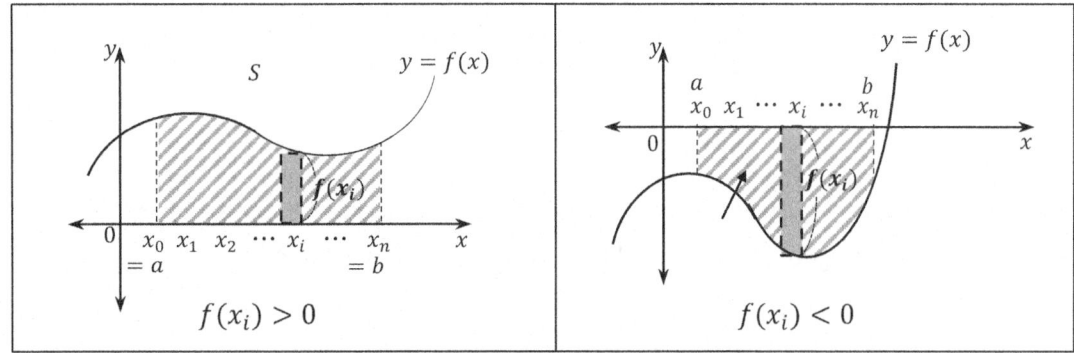

The area of each rectangle is $f(x_i)\Delta x$, where $\Delta x = x_i - x_{i-1}$.

The area of all rectangles under the curve from a to b is $\displaystyle\sum_{i=1}^{n} f(x_i)\Delta x$.

Therefore, the area of the region under the curve between $x = a$ and $x = b$ is

$$S = \lim_{n \to \infty} \sum_{i=1}^{n} f(x_i)\Delta x = \int_a^b f(x)dx$$

If the continuous function $y = f(x)$ is negative on the interval $a \leq x \leq b$ (the graph of $y = f(x)$ is below the x-axis), then $f(x_i)\Delta x \leq 0$.

$$\therefore \ \lim_{n \to \infty} \sum_{i=1}^{n} f(x_i)\Delta x \leq 0 \qquad \therefore \ \int_a^b f(x)dx \leq 0 \qquad \therefore \ S = -\int_a^b f(x)dx$$

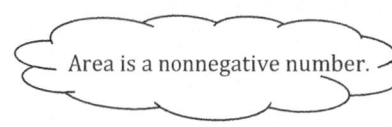

Area is a nonnegative number.

$$S_1 = \int_a^c f(x)dx, \quad S_2 = -\int_c^b f(x)dx$$

$$\therefore \ S_1 + S_2 = \int_a^c f(x)dx - \int_c^b f(x)dx = \int_a^c |f(x)|dx + \int_c^b |f(x)|dx$$

$$= \int_a^b |f(x)|dx$$

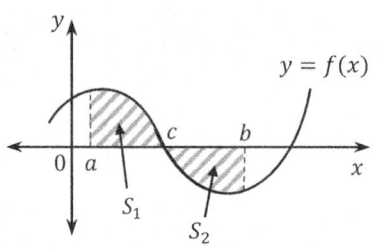

Note: (1) A region above the x-axis $\Rightarrow S = \int_a^b f(x)dx$

(2) A region below the x-axis $\Rightarrow S = -\int_a^b f(x)dx$

(3) In general, $S = S_1 + S_2 \Rightarrow S = \int_a^b |f(x)|dx$

Specially, $S_1 = S_2 \Rightarrow S = \int_a^b |f(x)|dx = 0$

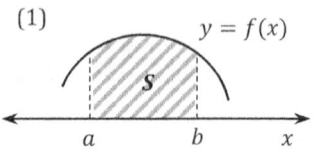

(1)

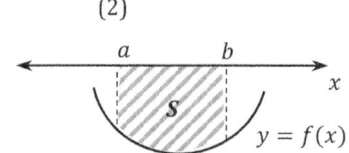

(2)

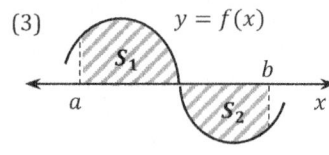

(3)

Example   Find the area of the region S under the curve $y = x^2 - 3x + 2$ between the x-axis and y-axis.

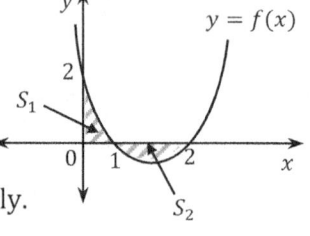

The region S is the shaded part in Figure.

One part of S is above the x-axis and the other part is below.

The areas of these two parts, S_1 and S_2, must be calculated separately.

$\therefore\ S = S_1 + S_2 = \int_0^2 |f(x)|dx = \int_0^1 f(x)dx - \int_1^2 f(x)dx = \int_0^1 (x^2 - 3x + 2)dx - \int_1^2 (x^2 - 3x + 2)dx$

$= \left[\frac{1}{3}x^3 - \frac{3}{2}x^2 + 2x\right]_0^1 - \left[\frac{1}{3}x^3 - \frac{3}{2}x^2 + 2x\right]_1^2$

$= \left(\frac{1}{3} - \frac{3}{2} + 2\right) - \left(\frac{1}{3}\cdot 8 - \frac{3}{2}\cdot 4 + 4\right) + \left(\frac{1}{3} - \frac{3}{2} + 2\right)$

$y = x^2 - 3x + 2$
$= (x - 1)(x - 2)$

$= 2\left(\frac{1}{3} - \frac{3}{2} + 2\right) - \left(\frac{8}{3} - 6 + 4\right) = 1$

2. A Region between a Curve in the xy-Plane and y-axis

Let $x = g(y)$ is a curve in the xy-pane and suppose g is continuous on the interval $c \leq y \leq d$.

Consider the region S bounded by the graph of $x = g(y)$, y-axis ($x = 0$), $y = c$, and $y = d$.

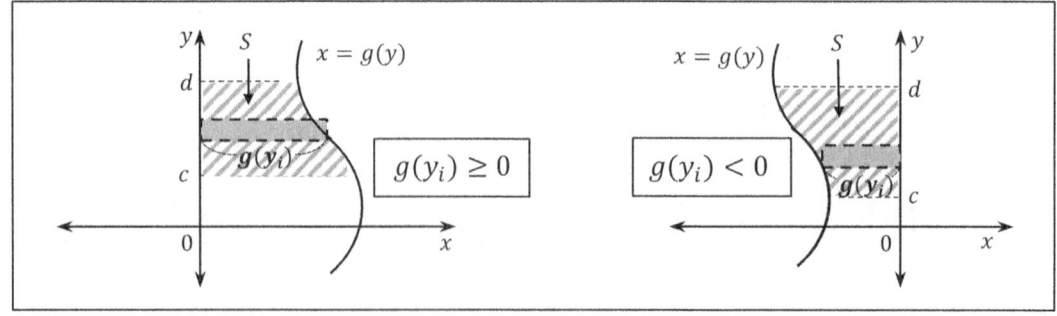

(i) When $x = g(y) \geq 0$,

$$S = \lim_{n \to \infty} \sum_{i=1}^{n} g(y_i)\Delta y = \int_c^d g(y)dy$$

(ii) When $x = g(y) < 0$,

Since $g(y_i)\Delta y < 0$, $\quad \lim_{n \to \infty} \sum_{i=1}^{n} g(y_i)\Delta y < 0 \quad \therefore \int_c^d g(y)dy < 0 \quad \therefore S = -\int_c^d g(y)dy$

That is, if the graph of the continuous function $x = g(y)$ is on the left of the y-axis on the

interval $c \leq y \leq d$, then $S = -\int_c^d g(y)dy$

Area is a nonnegative number.

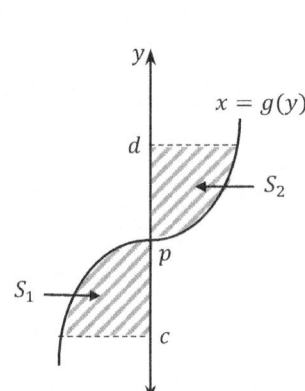

$$S_1 = -\int_c^p g(y)dy, \quad S_2 = \int_p^d g(y)dy$$

$$\therefore \ S_1 + S_2 = -\int_c^p g(y)dy + \int_p^d g(y)dy = \int_c^p |g(y)|dy + \int_p^d |g(y)|dy$$

$$= \int_c^d |g(y)|dy$$

Note: (1) A region on the right of the y-axis $\Rightarrow S = \int_c^d g(y)dy$

(2) A region on the left of the y-axis $\Rightarrow S = -\int_c^d g(y)dy$

(3) In general, $S = S_1 + S_2 \Rightarrow S = \int_c^d |g(y)|dy$

Specially, $S_1 = S_2 \Rightarrow S = \int_c^d |g(y)|dy = 0$

(1)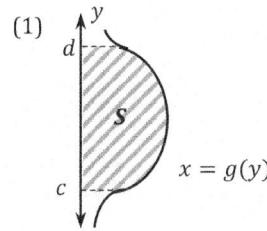

(2)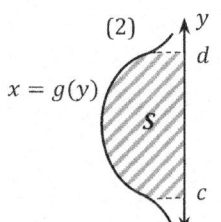

(3)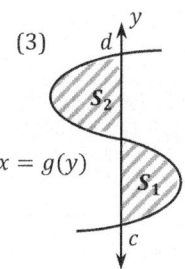

Example Determine the area of the region bounded by $y^2 = x$, $y = 2$, and y-axis.

(1) Using dx,

$\quad S = $ (The area of $\square\, OABC$) $-$ (The area of the region bounded by $y^2 = x$, x-axis, and $x = 4$)

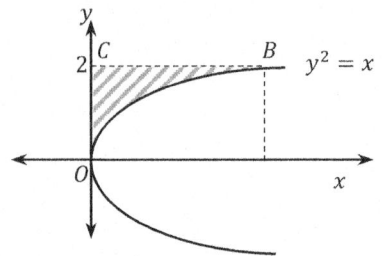

$$= 4 \cdot 2 - \int_0^4 y\,dx = 8 - \int_0^4 \sqrt{x}\,dx = 8 - \left[\frac{2}{3}x^{\frac{3}{2}}\right]_0^4$$

$$= 8 - \left(\frac{2}{3}4^{\frac{3}{2}}\right) = 8 - \frac{2}{3}\cdot 2^3 = 8 - \frac{16}{3} = \frac{8}{3}$$

(2) Using dy,

$$S = \int_0^2 x\,dy = \int_0^2 y^2\,dy = \left[\frac{1}{3}y^3\right]_0^2 = \frac{1}{3}\cdot 2^3 = \frac{8}{3}$$

7-2 The Area of a Region between Two Curves

Suppose that $f(x)$ and $g(x)$ are continuous on an interval $[a, b]$ such that $f(x) \geq g(x)$ for all $a \leq x \leq b$.

The area of the region bounded by $f(x)$ and $g(x)$ on $[a, b]$ is

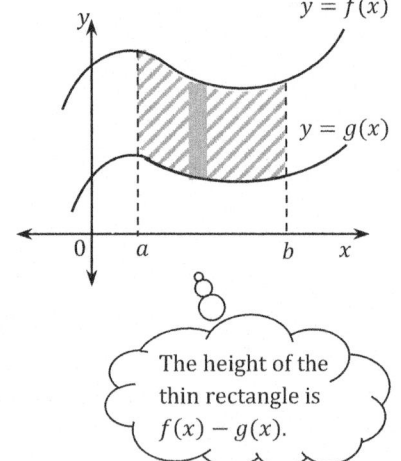

$$\int_a^b f(x)\,dx - \int_a^b g(x)\,dx = \int_a^b \{f(x) - g(x)\}\,dx.$$

If $g(x) > f(x)$, then the area is equal to $\displaystyle\int_a^b \{g(x) - f(x)\}\,dx.$

Above Function

Below Function

The height of the thin rectangle is $f(x) - g(x)$.

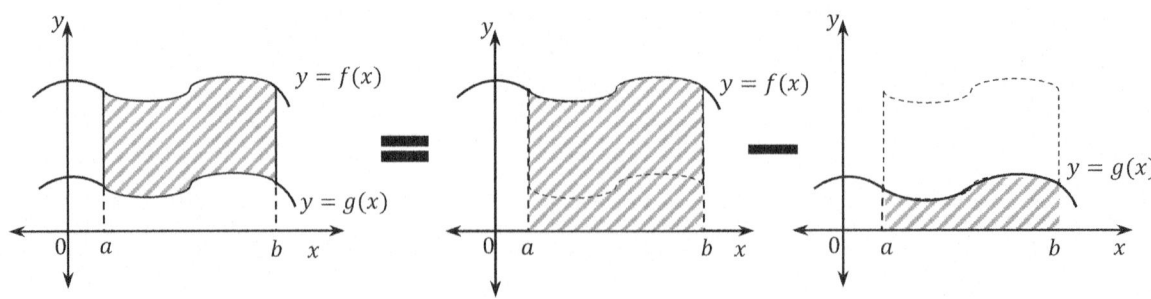

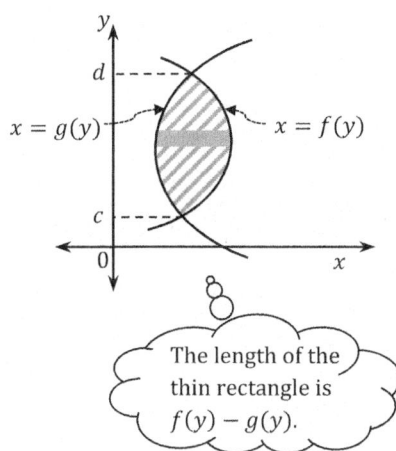

If $x = f(y)$ and $x = g(y)$ are continuous on an interval $[c, d]$

so that $f(y)$ is the function to the right of $g(y)$,

then the area of the region bounded by $x = f(y)$ and $x = g(y)$

on $[c, d]$ is $\displaystyle\int_c^d f(y)dy - \int_c^d g(y)dy = \int_a^b \{f(y) - g(y)\}dy$

Right Function

Left Function

The length of the thin rectangle is $f(y) - g(y)$.

Example Find the area of the region between the curves $y = x^2 - 1$ and $y = -x^2 + 2x + 3$.

First, we need to find the intersection points of the two curves.

To do this, solve $x^2 - 1 = -x^2 + 2x + 3$.

Then, $2x^2 - 2x - 4 = 0$; $x^2 - x - 2 = 0$; $(x - 2)(x + 1) = 0$

$\therefore\ x = 2$ or $x = -1$

Thus, the intersection points are $x = 2$ and $x = -1$.

The area of the shaded region is

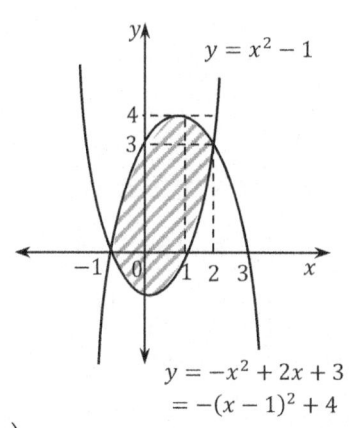

$$\int_{-1}^2 (-x^2 + 2x + 3)dx - \int_{-1}^2 (x^2 - 1)dx$$

$$= \int_{-1}^2 (-x^2 + 2x + 3 - x^2 + 1)dx$$

$$= \int_{-1}^2 (-2x^2 + 2x + 4)dx$$

$$= \left[-\frac{2}{3}x^3 + x^2 + 4x\right]_{-1}^2 = \left(-\frac{2}{3}\cdot 8 + 4 + 8\right) - \left(-\frac{2}{3}(-1) + 1 - 4\right)$$

$$= -\frac{18}{3} + 15 = \frac{27}{3} = 9$$

7-3 Volumes of Solids

1. Volumes of Solids with known Cross-Sections

By slicing a three-dimensional solid into thin cross-sectional slices, determining the volumes of the cross-sections, and calculating the sum of the volumes, the volume of the solid is defined.

Consider a solid which has cross-sections perpendicular to the x-axis along the interval $[a, b]$, and n intervals on the x-axis from a to b by parts, $a = x_0 < x_1 < x_2 < \cdots\cdots < x_n = b$ such that $x_i - x_{i-1} = \Delta x$.

Let $A(x)$ be the area of the cross-section at x, $a \le x \le b$.

Then, the volume of the cross section is $A(\bar{x}_i)\Delta x$, $x_{i-1} < \bar{x}_i < x_i$ and the volume V of the solid is given approximately by the Riemann sum $V \approx \sum_{i=1}^{n} A(\bar{x}_i)\Delta x$.

Thus, we obtain $V = \lim_{n \to \infty} \sum_{i=1}^{n} A(\bar{x}_i)\Delta x = \int_a^b f(x)dx$.

Example   A region lies between two planes perpendicular to the x-axis, one at $x = 0$ and one at $x = 1$. When the area $A(x_0)$ of the cross-section at $x = x_0$ is $3x_0 + 2$, find the volume of the region.

$$V = \lim_{n \to \infty} \sum_{i=1}^{n} A(\bar{x}_i)\Delta x = \int_0^1 A(x)dx = \int_0^1 (3x + 2)dx = \left[\frac{3}{2}x^2 + 2x\right]_0^1 = \frac{3}{2} + 2 = \frac{7}{2}$$

2. Volumes of Solids of Revolution

(1) Disk Method

1) About x-axis (Horizontal)

When a plane region, lying on the side of a fixed line is revolved about the line, we see a typical cross-section, *solid of revolution*.

The fixed line is called the *axis of the solid of revolution*.

Consider a plane region bounded by

$y = f(x)$, $x = a$, $x = b$ $(a < b)$, and the x-axis.

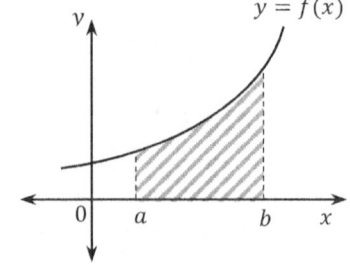

The area of the disk (circular cross-section) is $A(x) = \pi y^2 = \pi\{f(x)\}^2$ and

the volume of the disk is $A(x)\Delta x = \pi\{f(x)\}^2\Delta x$.

Thus, the volume of the solid of revolution is

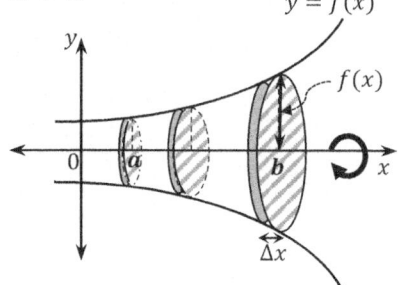

$$V = \lim_{n\to\infty}\sum_{i=1}^{n}A(\overline{x_i})\Delta x = \lim_{n\to\infty}\sum_{i=1}^{n}\pi\{f(x)\}^2\Delta x = \pi\int_{a}^{b}\{f(x)\}^2dx$$

Example   Rotate the region bounded by

(1) a curve $y = x^2$, the x-axis, $x = 1$, and $x = 2$

(2) a curve $y = e^x$, the x-axis, $x = -1$, and $x = 1$

about the x-axis. Find the volume of each solid of revolution.

(1) $V = \pi\displaystyle\int_{1}^{2}y^2dx = \pi\int_{1}^{2}(x^2)^2dx = \pi\int_{1}^{2}x^4dx$

$\qquad = \pi\left[\dfrac{1}{5}x^5\right]_{1}^{2} = \dfrac{\pi}{5}(2^5 - 1^5) = \dfrac{31}{5}\pi$

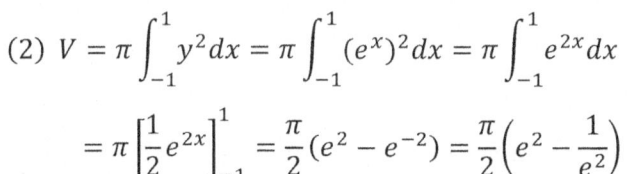

(2) $V = \pi\displaystyle\int_{-1}^{1}y^2dx = \pi\int_{-1}^{1}(e^x)^2dx = \pi\int_{-1}^{1}e^{2x}dx$

$\qquad = \pi\left[\dfrac{1}{2}e^{2x}\right]_{-1}^{1} = \dfrac{\pi}{2}(e^2 - e^{-2}) = \dfrac{\pi}{2}\left(e^2 - \dfrac{1}{e^2}\right)$

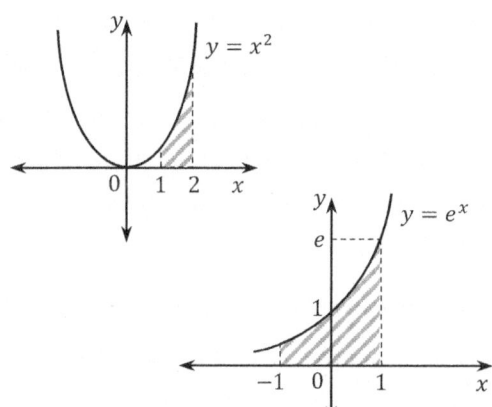

2) About y-axis (Vertical)

Similarly, the volume of the solid generated by revolving the region

bounded by $x = f(y)$, $y = a$, $y = b$ $(a < b)$, and the y-axis about the y-axis is

$$V = \lim_{n\to\infty}\sum_{i=1}^{n}A(\overline{y_i})\Delta y = \lim_{n\to\infty}\sum_{i=1}^{n}\pi\{f(y)\}^2\Delta y = \pi\int_{a}^{b}\{f(y)\}^2dy$$

The left and right boundary equations are expressed in terms of y.

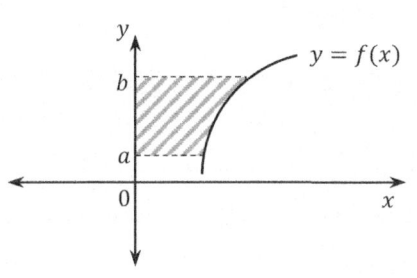

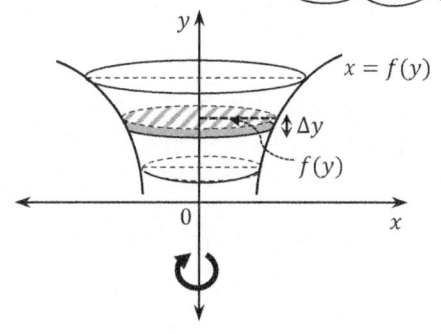

Example   Rotate the region bounded by $y = x + 1$, the y-axis, $y = 2$, and $y = 4$
about the y-axis. Find the volume of the solid of revolution.

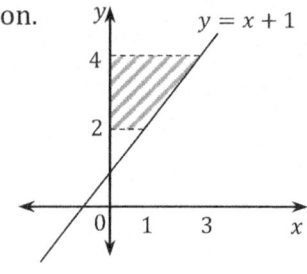

$$V = \pi \int_2^4 x^2 dy = \pi \int_2^4 (y - 1)^2 dy = \pi \int_2^4 (y^2 - 2y + 1)dy$$

$$= \pi \left[\frac{1}{3} y^3 - y^2 + y \right]_2^4 = \pi \left\{ \frac{1}{3}(4^3 - 2^3) - (4^2 - 2^2) + (4 - 2) \right\}$$

$$= \pi \left(\frac{56}{3} - 12 + 2 \right) = \frac{26}{3}\pi$$

Note:

Area bounded by the x-axis is $A(x) = \displaystyle\int_a^b ydx$

Area bounded by the y-axis is $A(y) = \displaystyle\int_a^b xdy$

Volume of the solid of revolution about the x-axis is $V = \pi \displaystyle\int_a^b y^2 dx$

Volume of the solid of revolution about the y-axis is $V = \pi \displaystyle\int_a^b x^2 dy$

(2) Washer Method

When slicing a solid of revolution results in thin circle shaped objects with holes in the middle, it is called *washer*.

Consider the area between two curves $y = f(x)$ and $y = g(x)$ is rotated around x-axis.

If the solid has a hole in the middle, then we compute the volume by subtracting the volume of the hole from the volume enclosed by the outer surface of the solid.

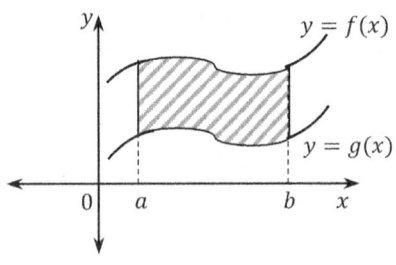 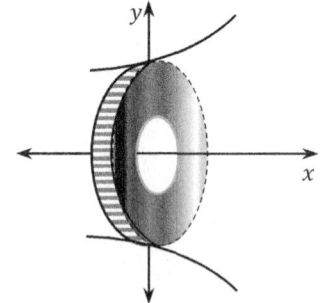

The volume of the solid generated by revolting the region bounded by $y = f(x)$ and $y = g(x)$

about the x-axis is $\boxed{V = \pi \displaystyle\int_a^b \{f(x)\}^2 dx - \pi \int_a^b \{g(x)\}^2 dx = \pi \int_a^b [\{f(x)\}^2 - \{g(x)\}^2]dx}$

Similarly, the volume of the solid generated by revolting

the region bounded by $x = f(y)$ and $x = g(y)$ about the y-axis is

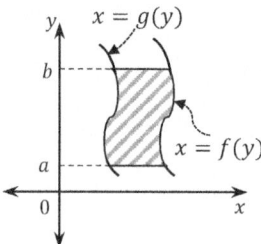

$$V = \pi \int_a^b \{f(y)\}^2 dy - \pi \int_a^b \{g(y)\}^2 dy = \pi \int_a^b [\{f(y)\}^2 - \{g(y)\}^2] dy$$

Example   When a region in the first quadrant is bounded by the curves $y = \sin x$, $y = \cos x$,

and the y-axis, find the volume of the solid generated by the rotating the region

about the line $y = -1$.

Since the region is rotated about the line $y = -1$, we have to consider the functions

$y = f(x) = \sin x + 1$ and $y = g(x) = \cos x + 1$.

Since $\sin x + 1 = \cos x + 1 \Rightarrow \sin x = \cos x$; $x = \dfrac{\pi}{4}$,

the intersection point of two curves is $x = \dfrac{\pi}{4}$.

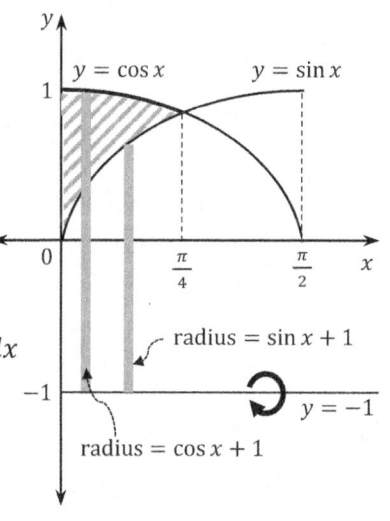

$$\therefore \quad V = \pi \int_0^{\frac{\pi}{4}} (\cos x + 1)^2 dx - \pi \int_0^{\frac{\pi}{4}} (\sin x + 1)^2 dx$$

$$= \pi \int_0^{\frac{\pi}{4}} [(\cos^2 x + 2\cos x + 1) - (\sin^2 x + 2\sin x + 1)] dx$$

$$= \pi \int_0^{\frac{\pi}{4}} (\cos^2 x - \sin^2 x + 2\cos x - 2\sin x) dx$$

$$= \pi \int_0^{\frac{\pi}{4}} (\cos 2x + 2\cos x - 2\sin x) dx$$

$$= \pi \left[\frac{1}{2} \sin 2x + 2\sin x + 2\cos x \right]_0^{\frac{\pi}{4}}$$

$$= \pi \left\{ \left(\frac{1}{2} \sin \frac{\pi}{2} + 2\sin \frac{\pi}{4} + 2\cos \frac{\pi}{4} \right) - \left(\frac{1}{2} \sin 0 + 2\sin 0 + 2\cos 0 \right) \right\}$$

$$= \pi \left\{ \left(\frac{1}{2} + 2 \cdot \frac{\sqrt{2}}{2} + 2 \cdot \frac{\sqrt{2}}{2} \right) - 2 \right\}$$

$$= \pi \left\{ \left(\frac{1}{2} + 2\sqrt{2} \right) - 2 \right\}$$

$$= \pi \left(2\sqrt{2} - \frac{3}{2} \right)$$

When a circular region is revolved about a line in its plane that does not intersect the circle, it sweeps out a torus (doughnut).

Consider a circle with center $(0, b)$ and radius a $(0 < a < b)$.

Then, the equation of the circle is $x^2 + (y - b)^2 = a^2$.

Since $0 < a < b$, the x-axis does not intersect the circle.

$x^2 + (y - b)^2 = a^2 \implies (y - b)^2 = a^2 - x^2$

$\therefore \ y - b = \pm\sqrt{a^2 - x^2}$

$\therefore \ y = b \pm \sqrt{a^2 - x^2}$

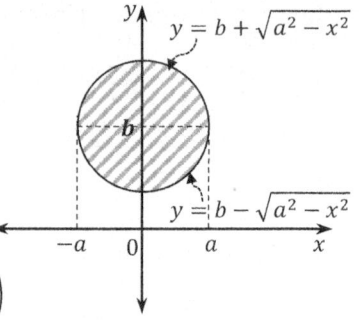

The volume of the above half solid generated by revolving the region bounded by $y = b + \sqrt{a^2 - x^2}$, $x = -a$, and $x = a$ about the x-axis is

$$V_1 = \pi \int_{-a}^{a} \left(b + \sqrt{a^2 - x^2}\right)^2 dx = 2\pi \int_0^a \left(b + \sqrt{a^2 - x^2}\right)^2 dx.$$

The volume of the below half solid generated by revolving the region bounded by $y = b - \sqrt{a^2 - x^2}$, $x = -a$, and $x = a$ about the x-axis is

$$V_2 = \pi \int_{-a}^{a} \left(b - \sqrt{a^2 - x^2}\right)^2 dx = 2\pi \int_0^a \left(b - \sqrt{a^2 - x^2}\right)^2 dx.$$

Thus, the volume of the whole circle generated by revolving about the x-axis is

$$V = V_1 - V_2 = 2\pi \int_0^a \left(b + \sqrt{a^2 - x^2}\right)^2 dx - 2\pi \int_0^a \left(b - \sqrt{a^2 - x^2}\right)^2 dx$$

$$= 2\pi \int_0^a \left\{\left(b + \sqrt{a^2 - x^2}\right)^2 - \left(b - \sqrt{a^2 - x^2}\right)^2\right\} dx = 2\pi \int_0^a \left(4b\sqrt{a^2 - x^2}\right) dx$$

$$= 8b\pi \int_0^a \left(\sqrt{a^2 - x^2}\right) dx$$

For $\int_0^a \left(\sqrt{a^2 - x^2}\right) dx,$

Let $x = a \sin \theta$ \quad Then, $dx = a \cos \theta \, d\theta$

$\theta = 0$ when $x = 0$

$\theta = \dfrac{\pi}{2}$ when $x = a$

$\therefore \ \int_0^a \left(\sqrt{a^2 - x^2}\right) dx = \int_0^{\frac{\pi}{2}} \left(\sqrt{a^2 - a^2 \sin^2 \theta}\right) (a \cos \theta \, d\theta) = a^2 \int_0^{\frac{\pi}{2}} \left(\sqrt{1 - \sin^2 \theta}\right) (\cos \theta \, d\theta)$

$$= a^2 \int_0^{\frac{\pi}{2}} \sqrt{\cos^2 \theta} \, (\cos \theta \, d\theta) = a^2 \int_0^{\frac{\pi}{2}} \cos^2 \theta \, d\theta = a^2 \int_0^{\frac{\pi}{2}} \left(\frac{1 + \cos 2\theta}{2} \right) d\theta$$

$$= \frac{a^2}{2} \int_0^{\frac{\pi}{2}} (1 + \cos 2\theta) d\theta = \frac{a^2}{2} \left[\theta + \frac{1}{2} \sin 2\theta \right]_0^{\frac{\pi}{2}} = \frac{a^2}{2} \left\{ \left(\frac{\pi}{2} + \frac{1}{2} \sin \left(2 \cdot \frac{\pi}{2} \right) \right) - \left(0 + \frac{1}{2} \sin 0 \right) \right\}$$

$$= \frac{a^2}{2} \cdot \frac{\pi}{2} = \frac{a^2 \pi}{4}$$

Therefore, $V = 8b\pi \cdot \dfrac{a^2 \pi}{4} = 2a^2 b\pi^2$

7-4 Arc Length

1. Distance, Velocity

Let an object P moves along a horizontal coordinate line so that its velocity $V(t)$ at time t is given by $y = V(t)$.

When $V(t)$ is continuous and $V(t) \geq 0$, the directed distance from $t = a$ to $t = b$ is $\int_a^b V(t)dt$.

Consider $V(t) = 4 - t$, $V(t) \geq 0$

Since the object P moves to the positive direction, the position S of the object from the origin to

$t = 4$ is $\int_0^4 V(t)dt = \int_0^4 (4 - t)dt = \left[4t - \frac{1}{2}t^2 \right]_0^4 = 4 \cdot 4 - \frac{1}{2} \cdot 4^2 = 8.$

Since $V(t) \leq 0$ when $t \geq 4$,

the object P moves to the negative direction (change its direction).

For example, the rate of change of S from $t = 4$ to $t = 6$ is

$$\int_4^6 V(t)dt = \int_4^6 (4 - t)dt = \left[4t - \frac{1}{2}t^2 \right]_4^6 = 4 \cdot (6 - 4) - \frac{1}{2} \cdot (6^2 - 4^2) = 8 - 10 = -2$$

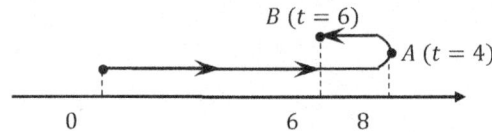

Note that the position S of the object from $t = 0$ to $t = 6$ is

$$\int_0^6 V(t)dt = \int_0^6 (4 - t)dt = \left[4t - \frac{1}{2}t^2 \right]_0^6 = 4 \cdot 6 - \frac{1}{2} \cdot 6^2 = 24 - 18 = 6$$

and also, the directed distance from $t = 0$ to $t = 6$ is $\overline{OA} + \overline{AB} = 8 + 2 = 10$

Thus, the directed distance from $t = a$ to $t = b$ is $\boxed{\int_a^b |V(t)|dt\, .}$

Note: Let $S(t)$ be the position of the object at time t (its distance from the origin on the x-axis).

Then, the change in position between $t = a$ to $t = b$ is $S(b) - S(a)$.

Since $S(t)$ is an anti-derivative of the velocity $V(t)$,

$$S(t) = S(a) + \int_a^b V(t)dt.$$

Similarly, since the velocity of an anti-derivative of the acceleration $A(t)$,

$$V(t) = V(a) + \int_a^b A(t)dt$$

Example   An object P moves along a horizontal coordinate line.

At the end of t seconds, its velocity from the origin is $V = 2\sin\pi t$ (feet/sec).

(1) Find the position of the object P at the end of 1 second and 2 seconds.

(2) Find the distance of P from $t = 0$ to $t = 2$.

(1) The position at the end of 1 second is

$$S(t) = 0 + \int_0^1 (2\sin\pi t)dt = \left[-\frac{2}{\pi}\cos\pi t\right]_0^1 = -\frac{2}{\pi}(\cos\pi - \cos 0) = -\frac{2}{\pi}(-1-1) = \frac{4}{\pi} \text{ feet.}$$

The position at the end of 2 seconds is

$$S(t) = 0 + \int_0^2 (2\sin\pi t)dt = \left[-\frac{2}{\pi}\cos\pi t\right]_0^2 = -\frac{2}{\pi}(\cos 2\pi - \cos 0) = -\frac{2}{\pi}(1-1) = 0 \text{ feet.}$$

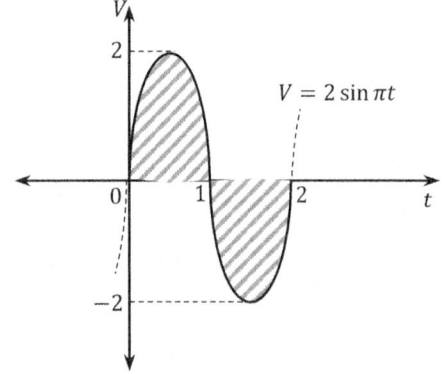

(2) The distance is

$$\int_0^2 |V(t)|dt = \int_0^2 |2\sin\pi t|dt = \int_0^1 (2\sin\pi t)dt + \int_1^2 (-2\sin\pi t)dt$$

$$= \left[-\frac{2}{\pi}\cos\pi t\right]_0^1 + \left[\frac{2}{\pi}\cos\pi t\right]_1^2 = -\frac{2}{\pi}(\cos\pi - \cos 0) + \frac{2}{\pi}(\cos 2\pi - \cos\pi)$$

$$= \frac{4}{\pi} + \frac{4}{\pi} = \frac{8}{\pi} \text{ feet}$$

2. Length of a Plane Curve

For a line segment $\overline{AB}$ with endpoints $A(x_1, y_1)$ and $B(x_2, y_2)$, the length of the segment is the distance between the points.

That is, $d(A,B) = \sqrt{(x_2 - x_1)^2 + (y_2 - y_1)^2}$

from the Pythagorean theorem.

$$d = \sqrt{(x_2 - x_1)^2 + (y_2 - y_1)^2}$$

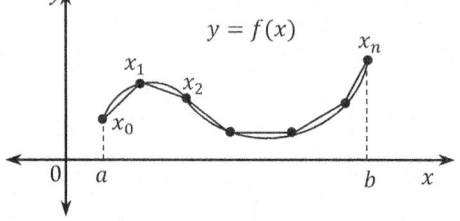

Now, find the length measured along a circle or some other curve.

If the curve of f is smooth, we can approximate the length of a portion of the curve with line segments, and that as the number of segments increases, and their lengths decreases, the limit of the sum of the lengths of the line segments will be the true arc length.

Suppose we divide the interval $[a, b]$ into n subintervals, each with length $\Delta x = \frac{b-a}{n}$, and endpoints $a = x_0 < x_1 < x_2 < \cdots\cdots < x_n = b$.

The length of a line segment connecting $(x_i, f(x_i))$ and $(x_{i+1}, f(x_{i+1}))$ is

$$\sqrt{(x_{i+1} - x_i)^2 + (f(x_{i+1}) - f(x_i))^2} = \sqrt{(\Delta x)^2 + (f(x_{i+1}) - f(x_i))^2}.$$

By the Mean Value theorem for derivatives, there is a number t_i in (x_i, x_{i+1}) such that

$$\frac{f(x_{i+1}) - f(x_i)}{\Delta x} = f'(t_i) \; ; \quad f'(t_i)\Delta x = f(x_{i+1}) - f(x_i)$$

$\therefore$ The length of the line segment is $\sqrt{(\Delta x)^2 + \{f'(t_i)\Delta x\}^2} = \sqrt{1 + \{f'(t_i)\}^2}\, \Delta x$.

Therefore, the length of the curve $y = f(x) \ (a \le x \le b)$ is

$$\lim_{n \to \infty} \sum_{i=0}^{n-1} \sqrt{1 + \{f'(t_i)\}^2}\, \Delta x = \lim_{n \to \infty} \sum_{i=1}^{n} \sqrt{1 + \{f'(t_i)\}^2}\, \Delta x = \int_a^b \sqrt{1 + \{f'(x)\}^2}\, dx = \int_a^b \sqrt{1 + \left(\frac{dy}{dx}\right)^2}\, dx$$

Similarly, if the curve is given by $x = g(y) \ (c \le y \le d)$, then the length of the curve is

$$\int_c^d \sqrt{1 + \left(\frac{dx}{dy}\right)^2}\, dy.$$

The length of an arc $\widehat{AB}$

If $A(a, c)$ and $B(b, d)$ are two points on the curve $F(x, y) = 0$, then the length l of the arc $\widehat{AB}$ is given by

$$l = \int_a^b \sqrt{1 + \left(\frac{dy}{dx}\right)^2}\, dx \quad \text{or} \quad l = \int_c^d \sqrt{1 + \left(\frac{dx}{dy}\right)^2}\, dy.$$

If A and B are the endpoints on a curve defined by the parametric equations $x = f(t)$, $y = g(t)$ $(a \le t \le b)$, then the length l of the arc $\widehat{AB}$ is given by

$$l = \int_a^b \sqrt{\left(\frac{dx}{dt}\right)^2 + \left(\frac{dy}{dt}\right)^2}\, dt\,.$$

For example, the length of the line segment from $A(1, 1)$ to $B(13, 6)$ is

$$l = \sqrt{(13 - 1)^2 + (6 - 1)^2} = \sqrt{12^2 + 5^2} = \sqrt{13^2} = 13 \quad \text{(By the Pythagorean theorem)}.$$

Using the formula;

The equation of the line connecting $A(1, 1)$ and $B(13, 6)$ is

$y = mx + b$ where $m = \frac{6-1}{13-1} = \frac{5}{12}$.

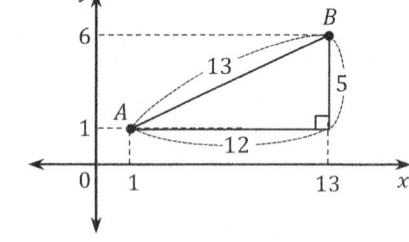

Since $(1, 1)$ lies on the line, $1 = \frac{5}{12} \cdot 1 + b$; $b = \frac{7}{12}$

$\therefore \quad y = \frac{5}{12}x + \frac{7}{12}$; $\frac{dy}{dx} = \frac{5}{12}$

By the formula,

$$l = \int_1^{13} \sqrt{1^2 + \left(\frac{5}{12}\right)^2}\, dx = \int_1^{13} \sqrt{\frac{12^2 + 5^2}{12^2}}\, dx = \int_1^{13} \sqrt{\frac{13^2}{12^2}}\, dx = \frac{13}{12}\int_1^{13} dx = \frac{13}{12}(13 - 1) = 13.$$

Example   Find the circumference of the circle $x^2 + y^2 = r^2$.

The equation of the circle in parametric form is $\begin{cases} x = r\cos t \\ y = r\sin t \end{cases}$, $0 \le t \le 2\pi$.

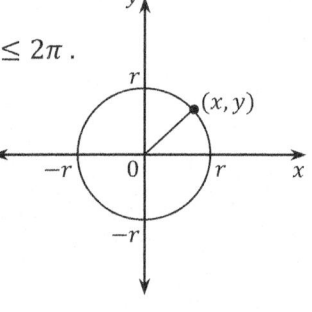

$\therefore \frac{dx}{dt} = -r\sin t$, $\frac{dy}{dt} = r\cos t$

By the formula,

$$l = \int_0^{2\pi} \sqrt{(-r\sin t)^2 + (r\cos t)^2}\, dt = \int_0^{2\pi} \sqrt{(r^2\sin^2 t + r^2\cos^2 t)}\, dt$$

$$= r\int_0^{2\pi} \sqrt{(\sin^2 t + \cos^2 t)}\, dt = r\int_0^{2\pi} dt = r[t]_0^{2\pi} = r(2\pi - 0) = 2\pi r$$

Exercises

#1 Sketch the region bounded by the graphs of the given equations and calculate the area S of the region.

(1) $y = x^2 - 4x + 3$, $x = 0$, and $y = 0$

(2) $y = x^3 - 3x^2 - x + 3$, x-axis between $x = -1$ and $x = 2$

(3) $y = x^3 + 4x^2 + 4x$, x-axis

(4) $y = x^3 - 3x^2 + 2x$, x-axis

(5) $y = \dfrac{1}{x}$ $(1 \le x \le e)$, x-axis

(6) $y = \sqrt{-x + 1}$, x-axis, y-axis, and $y = 2$

(7) $y^2 = 4 - x$, $x = 0$

(8) $y^2 = x + 1$, $y = 3$, and $x = 0$

(9) $1 + \log_{10} y = 4 \log_{10} x$, $x = 4$, x-axis

(10) $y = -\log x$ $(1 \le x \le e)$, x-axis

(11) $y = \log(x + 1)$, $x = 0$, $y = \log 3$, $y = -\log 3$

(12) $y = x \sin x$ $(1 \le x \le 2\pi)$, x-axis

(13) $y = \tan x$, $x = \dfrac{\pi}{3}$, $y = 0$

(14) $x = \cos y$ $(0 \le y \le \pi)$, $y = 0$, $y = \pi$, $x = 0$

(15) $\sqrt{x} + \sqrt{y} = 1$, $x = 0$, $y = 0$

(16) $y = |e^x - 1|$, $y = 0$, $y = -1$, $x = 1$

#2 Find the value.

(1) For the area S of the region between a curve $y = x(x - 1)(x - a)$, $a > 1$, and the x-axis, find the value of a so that the areas of parts of S are the same.

(2) For the area S of the region bounded by $y = (x - a) \sin x$ $(0 \le x \le \pi)$ and the x-axis, find the value of a so that the areas of parts of S are the same. $(0 < a < \pi)$

(3) When the area of the region bounded by $y^2 = 4 - ax$ $(a > 0)$ and the y-axis is $\dfrac{1}{3}$, find the value of a.

(4) When the area of the region bounded by $y = x(x - a)^2$ $(a < 0)$ and the x-axis is 12, find the value of a.

(5) When a curve $y = \sqrt{ax}$ divides the area of the region bounded by $y = \sqrt{x}$ and the x-axis into two equal parts, find the value of a. $(a > 0)$

(6) Find the value of a so that the area of the region bounded by $y = \sqrt{x + a}$, $x = 0$, and $y = 0$

is $\dfrac{5}{2}$.

(7) For a function $f(x) = e^x - 1$, let g be the inverse of f. When a is a positive constant,

find the value of $\displaystyle\int_0^a f(x)dx + \int_0^{f(a)} g(t)dt$.

(8) Let S_1 be the area of the region bounded

by a curve $y = x^2 - 4x + a$, the x-axis, and the y-axis, and

S_2 be the area of the region bounded by the curve y and the x-axis.

Find the area of a such that $S_1 : S_2 = 1 : 2$.

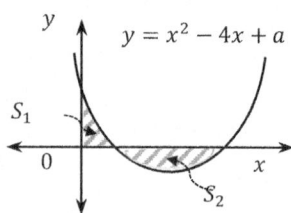

(9) Let S_1 be the area of the region bounded by a curve $f(x) = x^2 - 7x + 10$, the x-axis, and

the y-axis, and S_2 be the area of the region bounded by the curve $f(x)$ and the x-axis,

and S_3 be the area of the region bounded by the curve $f(x)$, the x-axis, and $x = a$ $(a > 5)$.

When $\{S_1, S_2, S_3\}$ forms an arithmetic sequence, find the value of $\displaystyle\int_0^a f(x)dx$.

(10) For a differentiable function $f(x)$, $f(x) \geq 0$ for all $x \geq 0$.

When the area of the region bounded by $y = f(x)$, the x-axis, the y-axis, and $x = a$ $(a > 0)$

is $F(a) = a \sin a + ab$, find the value of $f(a)$ such that $f(0) = 0$.

(11) For the area S_n of the region bounded by a curve $y = -\dfrac{1}{n(n+1)}x(x + 1)$ and the x-axis,

find the value of $\displaystyle\sum_{n=1}^{\infty} S_n$.

(12) Find the minimum value of the area of the region bounded by $y = x^3 + (x - a)^2 + 1$,

$x = 0$, $x = 1$, and $y = 0$.

#3 Find the area of the region bounded by the given curves.

(1) $y = 3x$, $y = x^2$ when $x > 0$.

(2) $y = \dfrac{1}{x}$, $y = x$, $y = \dfrac{1}{2}x$, $x > 0$

(3) $y = -x$, $y = -x^2 + 2x$

(4) $y = x$, $y = x^3 - 6x^2 + 9x$

(5) $y = x^2$, $y = x^3 - 2x$

(6) $y = 2x - x^2$, $y = x^4$

(7) $y = \dfrac{1}{3}x^2 + 1$, $y = -\dfrac{1}{3}x^2 + 3$, $x = 0$, $x = 3$

(8) $y = x^3 - 3x^2 + 2x$, $y = 0$, $x = 3$

(9) $y = x^2$, $y = x + 2$, $y = x$

(10) $y \geq x,\ x^2 - xy - xy^2 + y^3 \leq 0$

(11) $y = xe^{-x},\ y = e^{-2}x$

(12) $y = \sin x,\ y = 3,\ x = -\dfrac{\pi}{2},\ x = \pi$

(13) $y = \sin x,\ y = \sin 2x,\ x = 0,\ x = \pi$

(14) $y = x^2,\ y^2 = 27x$

(15) $y = -x + 2,\ y^2 = x$

(16) $y = x^2,\ y = \sqrt{x}$

(17) $x - y = 3,\ x = y^2 - y$

(18) $y = x^2,\ y = 4\sqrt{x} - 3,\ x = 0$

(19) $y = x,\ y = e^x,\ y = 1,\ y = 3$

(20) $y = e^x,\ y = ex,\ x = 0$

(21) $y = \log x,\ y = \dfrac{x}{e},\ y = 0$

(22) $y = \dfrac{1}{1+x^2},\ x = \sqrt{3},$ x-axis, y-axis

(23) $y = x^{2n-1},\ y = x^{2n+1}$

(24) $y = x^n,\ y^n = x\ (x \geq 0)$

#4 Find the area.

(1) Find the area of the region bounded by the curves $y = \sqrt{x+1}$, the normal line at $(3, 2)$ on the graph of y, and the x-axis.

(2) Find the area of the region bounded by the curves $y = x^3 - 3x^2 + 5$ and the tangent line at the local minimum value of the curve.

(3) For a curve $y = x^3 - x$, a line l passing through the origin is perpendicular to the tangent line at the origin to the curve.

Find the area of the region bounded by the curve and the line l.

(4) When the tangent line at $(2, 1)$ on the curve $y = x^3 + ax + b$ is $y = x + c$, find the area of the region bounded by the curve and the line.

(5) When $x = 2$ is the point of tangency for the curves $f(x) = x^3 - (a + 1)x^2 + ax$ and $g(x) = x^2 - ax$, find the area of the region bounded by the two curves.

(6) When a curve $y = x \log x + x + a$ and a line $y = 0$ intersect only at one point, find the area of the region bounded by the curve y, $x = 1$, and $y = 0$.

(7) When two curves $y = ax^2$ and $y = \log x$ intersect only at one point, find the area of the region bounded by the curves and the x-axis.

(8) Find the area of the region bounded by the curves $y = e^x$, a tangent line at $(0, 0)$ to the curve, a vertical line $x = -2$, and the x-axis.

(9) Find the area of the ellipse $\dfrac{x^2}{a^2} + \dfrac{y^2}{b^2} = 1$. $(a > 0, \ b > 0)$

(10) For a differentiable function $f(x)$ $(x \geq 0)$, $\displaystyle\int_a^x f(t)dt = \dfrac{4}{3}x\sqrt{x} - \dfrac{1}{2}x^2 + 1$.

Find the area of the region bounded by $y = f(x)$ and the x-axis.

(11) When a function $f(x) = \displaystyle\lim_{n \to \infty} \dfrac{x^{n+1} + ax^2 + b}{x^n + 1}$ $(x > 0)$ is differentiable at $x = 1$,

find the area of the region bounded by the curve $y = f(x)$, $x = 3$, and the x-axis.

(12) Find the area of the region bounded by the curve $y = x^2 - 2x + 7$ and the tangent lines at $(1, -3)$ to the curve.

(13) For a curve $f(x) = ax^2 + b$ $(x \geq 0)$, $g(x)$ is the inverse of $f(x)$. When $f(x)$ and $g(x)$ intersect at $x = 1$ and $x = 2$, find the area A of the region bounded by the curves, the x-axis, and the y-axis on $[0, 1]$, and the area B of the region bounded by the curves on $[1, 2]$.

#5 When the base of a solid is bounded by the circle $x^2 + y^2 = 16$,

(1) Find the volume of the solid with the base that has semicircular cross-sections perpendicular to the y-axis.

(2) Find the volume of the solid with the base that has semicircular cross-sections perpendicular to the x-axis.

#6 Find the volume.

(1) Let the base of a solid be the first quadrant plane region bounded by $4x^2 + 9y^2 = 36$, the x-axis, and the y-axis. When the cross-sections perpendicular to the x-axis are squares, find the volume of the solid.

(2) Let the base of a solid be the first quadrant plane region bounded by $y = 1 - \dfrac{x^2}{2}$, the x-axis, and the y-axis. Suppose that cross-sections perpendicular to the x-axis are squares. Find the volume of the solid.

(3) A solid has a base in the form of an ellipse with major axis 8 and minor axis 4. Find the volume if every section perpendicular to the major axis is an isosceles triangle with altitude 5.

(4) The base of a solid is the region between $f(x) = x^2 - 1$ and $g(x) = -x^2 + 1$, and its cross-sections perpendicular to the x-axis are equilateral triangles. When the solid has been truncated to show a triangular cross-section above $x = \dfrac{2}{3}$, find the volume of the solid.

#7 The base of a solid is the region between one arch of $y = \sin x$ $(0 \le x \le \pi)$ and the x-axis.

(1) Each cross-section perpendicular to the x-axis is a circle sitting on this base.

Find the volume of the solid.

(2) Each cross-section perpendicular to the x-axis is an equilateral triangle sitting on this base.

Find the volume of the solid.

#8 The base of a solid is the region bounded by $y = \sqrt{x}$, the x-axis, and $x = 4$.

(1) Find the volume of the solid with base that has square cross-sections perpendicular to the x-axis.

(2) Find the volume of the solid with base that has square cross-sections perpendicular to the y-axis.

(3) Find the volume of the solid with base that has rectangular cross-sections of height 2 that are perpendicular to the x-axis.

#9 A wedge is cut from a right circular cylinder of radius 3. The upper surface of the wedge is in a plane through a diameter of the circular base and makes an angle $45°$ with the base.

Find the volume of the wedge.

#10 For the region bounded by the graphs of the given equations, find the volume V of the solid generated by revolving the region about the x-axis.

(1) $y = 1 - x^2$, $y = 0$

(2) $y = x^2 + 1$, x-axis, $x = -1$, $x = 1$

(3) $y = \sqrt{x + 1}$, $x = 0$, $y = 0$

(4) $y = 1 - |x|$, $y = 0$

(5) $x^2 + 4y^2 = 4$, $y = 0$

(6) $y = x - 2\sqrt{x}$ $(0 \le x \le 4)$, $y = 0$

(7) $y = 1 + \sin\dfrac{x}{2}$, $x = 0$, $y = 0$, $x = 2\pi$

(8) $y = e^x - 1$, $x = 1$, $y = 0$

(9) $y = \log x$, $x = e$, $y = 0$

(10) $y = 4 - x^2$, $y = 2 - x$

(11) $y = x^2$, $y = 2 - x^2$

(12) $y = e^{-x}$, $x = 0$, $y = 0$, $x = 1$

(13) $y = e^x + e^{-x}$, $x = 0$, $y = 0$, $x = 1$

(14) $y = \tan x$, $x = \dfrac{\pi}{4}$, $y = 0$

(15) $y = \sin x$, $y = \sin 2x$ $\left(0 \le x \le \dfrac{\pi}{2}\right)$

(16) $y = \cos x + 1$, $x = \pi$, $x = -\pi$, $y = 0$

(17) $y = \sin x + \cos x$, $x = 0$, $x = \pi$, $y = 0$

(18) $y = x^2 - 1$, $y = x + 1$

(19) $y^2 = x + 2$, $y = x$

(20) $x^2 + (y - 4)^2 = 4$

(21) $y = \sqrt{9 - x^2}$, $y = \sqrt{1 - \dfrac{x^2}{9}}$, $x = -1$, $y = 2$

#11 For the region bounded by the graphs of the given equations, find the volume V of the solid generated by revolving the region about the y-axis.

(1) $y = 1 - x^2$, $y = 0$

(2) $y = \sqrt{x + 1}$, $x = 0$, $y = 0$

(3) $y = \sqrt{x + 1}$, $y = x + 1$

(4) $y = \sqrt[3]{x^2}$, $y = 1$

(5) $y = 1 - |x|$, $y = 0$

(6) $y = x^2$, $y = x + 2$

(7) $y^2 = x$, $y = x$

(8) $(x - 4)^2 + (y - 2)^2 = 1$

(9) $y = \log(x + 1)$, $x = 0$, $y = 1$

(10) $y = \log x$, $y = \log 2$, $x = 0$, $y = 0$

(11) $y = \log(2 - x)$, $x = 0$, $y = 0$

(12) $y = e^x$, $x = 0$, $y = e$

(13) $y = e^x - x - 1$, $x = 0$, $x = 1$

(14) $y = e^{-x^2}$ $(0 \le x \le 1)$, $x = 0$, $y = e^{-1}$

#12 Find the value.

(1) For the region bounded by $y = ax$ $(a > 0)$, $y = 0$, $x = 1$, and $x = 2$, the volume of the solid generated by revolving the region about x-axis is equal to 14π. Find the value of a.

(2) When the volume of the solid generated by revolving the region bounded by the ellipse $x^2 + ay^2 = 1$ $(a > 0)$ about the x-axis is equal to $\dfrac{\pi}{3}$, find the value of a.

(3) When a point $(1, 0)$ is the point of tangency of the curve $y = ax^4 + bx^2 + 1$, find the volume of the solid generated by revolving the region bounded by the curve y and $y = 0$ about the x-axis.

(4) Find the value of a at which the volume of the solid generated by revolving the region bounded by the curve $x^2 + a(a - 1)y^2 = a^2$ $(a > 1)$ about the x-axis has minimum value.

(5) When the volume of the solid generated by revolving the region bounded by the curve $y^2 = x + a^2$ $(a > 0)$ and the y-axis is equal to $\frac{16}{15}\pi$, find the value of a.

#13 Prove it.

(1) Prove the volume of a sphere with radius r is $\frac{4}{3}\pi r^3$.

(2) Prove the volume of a right circular cone with radius r and height h is $\frac{1}{3}\pi r^2 h$.

(r, h; non-zero real numbers.)

#14 Find the volume of the solid generated by rotating the following region about the given line.

(1) Rotating the region bounded by a curve $y = 2 - x^2$ and the x-axis about the line $y = -1$.

(2) Rotating the region bounded by a curve $y = 2 - x^2$ and the x-axis about the line $y = 2$.

#15 An object P is moving along a coordinate line subject to the indicated velocity V (in feet per second). Find the directed distance S (in feet) during the given time interval.

(1) $V = 3t^2 - 9t + 6$ (from $t = 0$ to $t = 3$)

(2) $V = e^t$ (from $t = 0$ to $t = 1$)

(3) $V = (t - 1)e^{-t}$ (from $t = 0$ to $t = 2$)

#16 When a ball is thrown upward from an initial height of 10 feet with a velocity of 20 feet per second, its velocity t seconds later is given by $V = 20 - 10t$.

(1) Find its height from ground level 3 seconds later.

(2) Find the maximum height that the ball reaches from the ground.

(3) Find the total distance during the first 3 seconds.

#17 A point $P(x, y)$ moves in the coordinate plane, starting at the origin.

When the velocity vector of the point at time t is given by $\mathbf{v}(t) = \langle e^t + 1, \ e^{-t} - 1 \rangle$, find the coordinates of P at $t = 1$.

#18 An object is moving along a coordinate line with the initial velocity 30 feet/sec. If the object decreases speed at $\frac{1}{2}$ feet/sec, find the distance from the starting to stopping point.

#19 A tank in the shape of a right circular cone is being filled with water. When the height of the tank is h, the area of its top is $S(h) = 4\pi h^2$. If the velocity of filling the tank is given by $V(t) = 2t(3 - t)$, find the maximum amount of water over the edge of the tank.

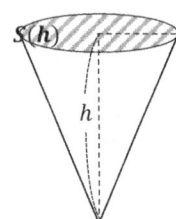

#20 A water tank in the shape of a rectangle with the surface area of 3 feet^2 is draining at a speed of $V(t) = 2t$ at t seconds. Find the amount of water that has drained during the first 10 seconds.

#21 When a certain object shoots straight up from the ground level with an initial velocity of a feet per second ($a > 0$), the velocity after t seconds is given by $V(t) = a - 10t$ feet/sec. What velocity will cause the object to reach its maximum height more than 20 feet?

#22 Starting from rest, an object is moving along a coordinate line with an initial velocity
$V(t) = \frac{3}{4}t^2 - \frac{1}{2}t + \frac{1}{2}$ (feet/sec) for the first 6 feet, and then travels at a constant speed.
Find the directed distance during the first 11 seconds.

#23 Find the length l of the indicated curve.
  (1) $y = 2x + 1$ $(0 \le x \le 1)$
  (2) $x = 6t^2$, $y = t^3 - 12t$ $(0 \le t \le 1)$
  (3) $x = t - \sin t$, $y = 1 - \cos t$ $(0 \le t \le 2\pi)$
  (4) $y = x\sqrt{x}$ $(0 \le x \le 2)$
  (5) $y = \frac{e^x + e^{-x}}{2}$ $(-1 \le x \le 1)$
  (6) $x = \log t$, $y = \frac{1}{2}\left(t + \frac{1}{t}\right)$ $\left(\frac{1}{e} \le t \le e\right)$
  (7) $x = e^{-t}\cos t$, $y = e^{-t}\sin t$ $(0 \le t \le \pi)$
  (8) $y = \int_0^x \frac{x - t}{\cos^2 t}\,dt$ $\left(0 \le x \le \frac{\pi}{6}\right)$

#24 When the length of the curve $y = f(x)$ from $(0, 1)$ to (x, y) on the curve is given by $l = e^{2x} + y - 1$, find the slope of the tangent line at $x = 2$ on the curve.
  ($y = f(x)$ is a differentiable function defined on $x \ge 0$.)

Chapter 1. The Concept of Limits

#1 (1) $\lim\limits_{n\to\infty} \dfrac{2}{n} = 0$ (2) $\lim\limits_{n\to\infty} \dfrac{n-1}{n+1} = 1$ (3) No limit exists.   (4) No limit exists.

(5) $\lim\limits_{n\to\infty}\left(1 - \dfrac{1}{10^n}\right) = 1$ (6) No limit exists.   (7) $\lim\limits_{n\to\infty}\left\{\log_2 \dfrac{n}{n+1}\right\} = 0$ (8) No limit exists.

(9) The limit is 0.   (10) No limit exists.

#2 (1) $\dfrac{3}{5}$ (2) $\dfrac{6}{5}$ (3) 0   (4) No limit exists.   (5) 0   (6) $\dfrac{3}{2}$ (7) $-\infty$ (8) 1   (9) $\dfrac{1}{2}$

(10) $\dfrac{1}{3}$ (11) $\dfrac{2}{3}$ (12) 0   (13) ∞ (14) $\dfrac{1}{2}$ (15) $-\infty$ (16) ∞ (17) 1   (18) $\dfrac{\sqrt{2}}{2}$

(19) a (20) 2   (21) 55   (22) $-\dfrac{2014}{2013}$ (23) 2

#3 (1) i) When $|r| < 1$, $\lim\limits_{n\to\infty} \dfrac{1}{1+r^n} = 1$

ii) When $r = 1$, $\lim\limits_{n\to\infty} \dfrac{1}{1+r^n} = \dfrac{1}{2}$

iii) When $|r| > 1$, $\lim\limits_{n\to\infty} \dfrac{1}{1+r^n} = 0$

(2) i) When $0 < r < 1$, $\lim\limits_{n\to\infty} \dfrac{r^{n+1}-1}{r^n+1} = -1$

ii) When $r = 1$, $\lim\limits_{n\to\infty} \dfrac{r^{n+1}-1}{r^n+1} = 0$

iii) When $r > 1$, $\lim\limits_{n\to\infty} \dfrac{r^{n+1}-1}{r^n+1} = r$

#4 (1) 10   (2) $\dfrac{5}{2}$ (3) $\dfrac{7}{2}$ (4) $\dfrac{2\sqrt{6}}{9}$ (5) 3   (6) 15   (7) 8   (8) $\dfrac{1}{5}$ (9) $-2 \le a \le 2$

(10) 0   (11) 8   (12) 4   (13) $a = \dfrac{\sqrt{2}}{2}$, $\lim\limits_{n\to\infty} a_n = \dfrac{-3\sqrt{2}}{8}$ (14) 25   (15) 6   (16) 6

(17) $\dfrac{1}{3}$ (18) -1

#5 (1) 1) $a_n = -5\left(\dfrac{1}{2}\right)^{n-1} + 6$ 2) 6   (2) 0   (3) 5   (4) 1) 0   2) 0

(5) 20   (6) 5   (7) 6   (8) 10   (9) $\dfrac{1}{3}$ (10) 2

#6 (1) True   (2) False   (3) False   (4) True   (5) False   (6) True   (7) False   (8) False

(9) True   (10) True   (11) i) True   ii) True   iii) False

#7 $\frac{4}{3}$

#8 (1) $a_{n+1} < \frac{1}{8}a_n$ (2) 0   (3) $\frac{1}{2}$

#9 $\frac{1}{2}$

#10 (1) $\log 2$ (2) ∞ (Diverge)   (3) $\frac{1}{2}$ (4) -1 (5) $-\frac{1}{5}$

#11 (1) The series is convergent.   (2) The series is divergent.

#12 (1) 2 (2) $\frac{2+\sqrt{3}}{2}$ (3) 1 (4) 6 (5) $\frac{15}{7}$ (6) $\frac{4}{7}$ (7) $\frac{32}{3}$ (8) $-\frac{1}{3}$ (9) $\frac{\sqrt{3}}{2}$ (10) 1

#13 (1) $\frac{1}{2}$ (2) 1) $n = 10$ 2) $a = 1$ (3) 1   (4) 1   (5) 1   (6) $-\frac{2}{3}$ (7) 4

   (8) $\frac{(24)^3}{(13)^3-(5)^3} \approx 6.672$ (9) $\frac{1}{2}$ (10) $\frac{1}{2}$ (11) $\frac{3}{4}S$ (12) $1 < x < 2$ or $-2 < x < -1$

   (13) $\frac{1}{2} < x \leq \frac{2}{3}$ (14) $\frac{1}{16}$ (15) $\frac{1}{9}$ (16) 20   (17) 24   (18) 3   (19) 8   (20) $\frac{16}{3}$

   (21) $n \geq 8$ (22) $\frac{3}{4}$ (23) 3   (24) $\frac{9}{4}$

#14 (1) True (2) False (3) False (4) True (5) False (6) False (7) True (8) True (9) True

#15 (1) True (2) True (3) False (4) True (5) True (6) False (7) False

Chapter 2. Limits of Functions and Continuity

#1 (1) 12   (2) 15   (3) $\frac{4}{3}$ (4) 2   (5) 1   (6) 1   (7) 1   (8) 3   (9) ∞ (10) ∞ (11) $-\infty$ (12) 7

(13) $3\frac{1}{2}$ (14) $2a^2$ (15) $\frac{1}{6}$ (16) 0   (17) 12   (18) $\frac{1}{3}$ (19) 0   (20) $\frac{5}{2}$ (21) ∞ (22) 1

(23) -1 (24) ∞ (25) 1   (26) -1 (27) $-\frac{1}{2}$ (28) $-\frac{1}{2}$

#2 (1) -6 (2) 6   (3) 16   (4) 2   (5) e

#3 (1) $a = 3$, $b = -3$ (2) $a = 3$, $b = -4$ (3) $a = 1$, $b = -\frac{1}{6}$

#4 $f(x) = 2x^2 - 5x + 2$

#5 (1) $\lim\limits_{x\to 0}\frac{1}{x}$ does not exist.   (2) $\lim\limits_{x\to\infty}\sin\frac{1}{x}$ exists.   (3) $\lim\limits_{x\to 1}\frac{|x^2-1|}{|x-1|}$ exists.

(4) $\lim\limits_{x\to\infty}\left(\sqrt{x^2-x}-x\right)$ exists.

#6 (1) False   (2) True   (3) True   (4) True   (5) False   (6) True   (7) False   (8) True   (9) False

#7 (1) 0   (2) $\frac{3}{2}$ (3) $-\frac{1}{4}$ (4) 3   (5) 3   (6) 2   (7) 6   (8) 4   (9) 2   (10) 7

(11) 0   (12) 1   (13) 2   (14) 12   (15) 0   (16) 3

#8 (1) The limit exists.   (2) The limit exists.   (3) The limit exists.

#9 (1) $\frac{3}{2}$ (2) $\frac{\pi}{180}$ (3) -3 (4) $\frac{4}{3}$ (5) 8   (6) $-\frac{1}{\pi}$ (7) $\frac{1}{4\pi}$ (8) 3   (9) 1   (10) 3   (11) $\log 3$

(12) $\log 6$ (13) 1   (14) $\log_e a$ (15) 1   (16) 2   (17) $-\frac{\pi}{2}$ (18) 0   (19) e (20) 2   (21) 385

(22) 1   (23) $\sqrt{e}$ (24) e (25) 1   (26) $\frac{1}{2}$ (27) π (28) 1   (29) 3   (30) $\sqrt{e}$ (31) $\frac{1}{e}$

#10 (1) $a = \frac{1}{2}$, $b = 0$ (2) $a = 2$, $b = 1$ (3) $a = 2$, $b = 0$

(4) $a = 3$, $b = 5$ (5) $a = 1$, $b = 2$ (6) $a = 2$, $b = 1$ (7) $a = 4$, $b = -4\pi$

#11 (1) $\frac{\pi}{2}$ (2) 1   (3) $-\frac{1}{3}$ (4) $\frac{4}{3}$ (5) $\frac{2}{5}$ (6) 4   (7) 1   (8) 4

#12 (1) False   (2) True   (3) True

#13 (1) True   (2) False   (3) True

#14 (1) False   (2) False   (3) True

#15 (1) True   (2) True   (3) False

#16 (1) 1   (2) 2   (3) 0

#17 (1) $x = 1$ (2) $x = 0$

#18 (1) $f(x)$ is discontinuous at $x = 2$.

(2) $f(x)$ is continuous at all real numbers in $[0, 1]$ except $x = \dfrac{1}{2}$.

(3) $f(x)$ is discontinuous at $x = 0$.

(4) $f(x)$ is discontinuous at $x = 0$.

#19 (1) $y = f(x)$ is continuous at all x in $(-\infty, \infty)$.

(2) $f(x)$ is discontinuous at $-x^2 + 5 = 1$, $-x^2 + 5 = -1$.

(3) $f(x)$ is discontinuous at $x = 1$.

(4) $f(x)$ is discontinuous at $x = \pm 1$.

#20 (1) (2) See the Solutions Manual.

#21 (1) $a = 2\sqrt{2}$, $b = 4$ (2) $a = 1$, $b = 0$ (3) $a = -2$, $b = 1$

(4) $a = -1$, $b = 2$ (5) $a = 3$, $b = 2$

#22 (1) $a = 2$ (2) $a = \dfrac{3}{2}$ (3) $a = \dfrac{7}{4}$ (4) $a = 1$

#23 (1) 0   (2) 90   (3) 0   (4) 1

#24 See the Solutions Manual

#25 (1) False   (2) True   (3) True

#26 (1) False   (2) True   (3) False   (4) True   (5) False   (6) False

Chapter 3. The Derivative

#1 (1) 2   (2) 2   (3) $\frac{1}{6}$ (4) $b^2 + ab + a^2$

#2 $\frac{b-a}{c-b}$

#3 $y - f(x_0) = 2x_0(x - x_0)$

#4 -10

#5 At $x = 2$, $f'(x)$ does not exist since the denominator is zero.

    At $x = 1$, $f'(x) = -1$

    At $x = 3$, $f'(x) = -1$

#6 $f'(x) = \frac{17}{(3x+4)^2}$ At $x = -\frac{4}{3}$, the derivative does not exist sine the denominator is zero.

#7 (1) $f(x)$ is not differentiable at $x = 0$.

    (2) $f(x)$ is differentiable at $x = 0$.

    (3) $f(x)$ is not differentiable at $x = 0$.

#8 (1) (2) (3) See Solutions Manual

#9 (1) 0   (2) 20   (3) i) -1 ii) $\frac{3}{2}$ (4) 6   (5) 3   (6) 2   (7) -2 (8) 6

#10 (1) $\frac{1}{\sqrt{2x+1}}$ (2) $\frac{3x^2+2x+1}{3\sqrt[3]{(x+1)^2(x^2+1)^2}}$ (3) $\frac{2x^3-x+2}{x^3}$ (4) $\frac{x^2-2x-3}{(x^2+3)^2}$ (5) $(x+1)^3(x^2-1)(7x-3)$

    (6) $\frac{3x^2(-x^2+1)}{(x^2+1)^4}$ (7) $\frac{1}{4\sqrt{x}\,(\sqrt{\sqrt{x}+1})}$ (8) $\frac{3x-5}{2\sqrt{x-4}}$ (9) $\frac{-3x-2}{(x^2+1)\sqrt{x^2+1}}$

#11 (1) $a = 0$, $b = -2$ (2) $a = -7$, $b = 6$ (3) $a = 4$, $b = 1$ (4) $a = -\frac{3}{2}$, $b = \frac{7}{2}$

    (5) $a = 3$, $b = \frac{1}{3}$ (6) $a = -1$, $b = 3$

#12 (1) $\frac{3}{2y}$ $(y \neq 0)$ (2) $\frac{2x}{3y^2}$ $(y \neq 0)$ (3) $-\frac{\sqrt{y}}{\sqrt{x}}$ $(x \neq 0)$ (4) $\frac{-x^2+y}{y^2-x}$ $(y^2 \neq x)$ (5) $\frac{2\sqrt{y+1}}{3y+2}$ (6) $\frac{x+2}{2}$

#13 (1) 45   (2) $3x$ (3) 0   (4) 2   (5) -5 (6) 2   (7) $\frac{10(x+\sqrt{1+x^2})^{10}}{\sqrt{1+x^2}}$ (8) -4 (9) $\frac{1}{5}$ (10) 5

    (11) 2   (12) $\frac{1}{2}$ (13) 32   (14) 6   (15) 20   (16) 23   (17) -1 (18) 5

#14 (1) True (2) True (3) False

#15 (1) True (2) False (3) True

#16 (1) $\dfrac{-x(\cos\sqrt{1-x^2})}{\sqrt{1-x^2}}$ (2) $-(\cos x)\sin(\sin x)$ (3) $-\dfrac{\pi}{180}\sin x^\circ$ (4) $4x\sin 2x + (4x^2+2)\cos 2x$

(5) $3\sin^2 x \cos 4x$ (6) $3(\sec x + \tan x)^3 \sec x$ (7) $2\pi \sin 2(2\pi x - a)$

(8) $6\sec^3(2x+5)\tan(2x+5)$ (9) $3e^x$ (10) $2(\log 3)\cdot 3^x$ (11) $e^x(1+x)$

(12) $e^x(\sin x + \cos x)$ (13) $\dfrac{e^x}{(e^x+1)^2}$ (14) $\dfrac{1-x\log 3}{3^x}$ (15) $3(x^2+e^x)^2(2x+e^x)$

(16) $\dfrac{1}{x}+1$ (17) $\log x$ (18) $\dfrac{3}{x\log 2}-3x^2$ (19) $e^x\left(\log x + \dfrac{1}{x}\right)$ (20) $\dfrac{1}{e^x}\left(\dfrac{1}{x}-\log x\right)$

(21) $\dfrac{3(\log_2 x)^2}{x\log 2}$ (22) $\sec x$ (23) $\dfrac{1}{\sqrt{x^2+1}}$ (24) $\dfrac{2}{x^2-1}$ (25) $(\log a)a^{\sin x}\cdot\cos x$

(26) $(\log 2)2^x \cos 2^x$ (27) $e^{x^2}(2x\sin x + \cos x)$ (28) $\dfrac{2x}{(x^2+1)\log 10}$

(29) $\log(x^2+1)+\dfrac{2x^2}{x^2+1}$ (30) $e^x\{\log(\sin x)+\cot x\}$ (31) $e^{x^x}\cdot x^x\cdot(\log x+1)$

#17 (1) $\dfrac{1}{\cos y}=\dfrac{1}{\pm\sqrt{1-x^2}}$ (2) $-\dfrac{1}{\sin y}=-\dfrac{1}{\sqrt{1-x^2}}$ (3) $-\dfrac{\cos x}{\cos y}$ (4) $\dfrac{\cos x}{\sin y}$ (5) $-\dfrac{\sin x\cos y}{\cos x\sin y}$

(6) $-\tan\theta$ (7) $-\dfrac{2}{3}\cot t$ (8) $\dfrac{y^2-xy\log y}{x^2-xy\log x}$ (9) $\dfrac{t^2-1}{2t}$

#18 (1) $\dfrac{1}{(x^2+1)\sqrt{x^2+1}}$ (2) $3(2\cos x\sin^2 x - \cos^3 x)$ (3) $2e^{x^2}(2x^2+1)$ (4) $\dfrac{-\log x+2}{x(\log x)^3}$

#19 (1) $f'(\pi)=-e^{a\pi}$, $f''(\pi)=-2ae^{a\pi}$ (2) $a=1$

#20 (1) -2 (2) 3 (3) $3e$ (4) 1 (5) $-\dfrac{1}{2}e^{\frac{\pi}{2}}$ (6) $x>e^{-2}$ (7) $\dfrac{2}{e}$

(8) $ae^a(2-a)$ (9) $n+1$ (10) $-\dfrac{1}{2}$ (11) -26 (12) $16e$ (13) $\dfrac{1}{5}$

(14) $\dfrac{1}{2}$ (15) $\dfrac{\sqrt{3}}{3}$ (16) 1 (17) $\dfrac{2}{e^a}$ (18) $\dfrac{1}{2}$

(19) At $(0,0)$, $-\dfrac{1}{3}$; At $(0,3)$, $\dfrac{1}{3}$ (20) 2 (21) $-\dfrac{1}{2}$ (22) $\dfrac{6+\ln 3}{\sqrt{4+(\ln 3)^2}}$

#21 (1) $-\dfrac{2}{\pi}$ (2) $\dfrac{1}{\pi\log 2}$ (3) $\dfrac{5}{3}$ (4) 3 (5) $\dfrac{11}{4}$

Chapter 4. Applications of the Derivative

#1 (1) $a = -1$, $b = 3$, $c = 1$ (2) $a = -\frac{4}{3}$, $b = 6$, $c = -\frac{14}{3}$ (3) $a = -4$, $b = 7$, $c = -2$

(4) $a = b = c = 1$

#2 (1) $y = 4x - 4$ (2) $y = 2x + 3$ (3) $\frac{1}{2e}x$ (4) $x_1 x + y_1 y = r^2$ (5) $y_1 y = 2p(x + x_1)$

(6) $\frac{x_1 x}{a^2} + \frac{y_1 y}{b^2} = 1$ (7) $y = -x + 2\pi$ (8) $y = -\frac{5}{8}x + 1$ (9) $y = 2x - 1$

(10) $y = -\frac{\sqrt{3}}{3}x + \frac{1}{2}$ (11) $y = 3x - \frac{25}{4}$

#3 (1) $y = -\frac{1}{3}x - \frac{4}{3}$ (2) $y = \frac{1}{3}x + \frac{5}{3}$ (3) $y = -\frac{1}{e^2}x + \frac{3}{e^2} + e$ (4) $y = -2x + 2$

(5) $y = -4x + 13$ (6) $y = -\frac{1}{4}x + \frac{\pi}{16} + 1$

#4 (1) $a = \frac{1}{2e}$ (2) $a = 2$, $a = -2$, or $a = \frac{5}{2}$ (3) $a = \frac{7}{48}$ (4) $a = \log 2$ (5) $y = -x + 1$

(6) $a = \frac{9}{4e^2}$ (7) $\frac{1}{2}$ (8) $= 5x - 6\sqrt{3}$, $y = 5x + 6\sqrt{3}$ (9) 8

(10) $r = \frac{\sqrt{17}}{4}$ (11) 16 (12) $a = 2\log 4$

#5 $y = \frac{1}{2}x + 1$

#6 (1) $c = 1$ (2) $c = \frac{3}{2}$ (3) $c = e - 1$

#7 $c = 0$

#8 (1) $9\log 3$ (2) 1

#9 (1) (2) See Solutions Manual

#10 (1) $a\cos a - \sin a$ (2) 0 (3) $-\frac{1}{24}$ (4) $\frac{1}{2}$ (5) 0 (6) ∞ (7) $\log a - \log b$ (8) e (9) $\frac{1}{\sqrt{e}}$

#11 (1) (2) (3) See Solutions Manual

#12 (1) (2) See Solutions Manual

#13 (1) $0 \le a \le 3$ (2) $a \le 0$ (3) $a = 4$ (4) $a \ge 1$ (5) $0 \le a \le 1$ (6) $-\sqrt{2} \le a \le \sqrt{2}$

#14 (1) Local maximum value is 0 at $x = -1$ and local minimum value is -4 at $x = 1$.

(2) Local maximum value is 13 at $x = 2$ and local minimum value is -19 at $x = -2$.

(3) There is no local maximum value.

(4) Local maximum value is 41 at $x = -3$ and local minimum value is -84 at $x = 2$.

(5) Local maximum value is 1 at $x = 2$ and local minimum value is 0 at $x = 1$.

(6) Local maximum value is $\frac{43}{6}$ at $x = 1$, local minimum value is $\frac{11}{6}$ at $x = -1$, and

local minimum value is $\frac{19}{3}$ at $x = 2$.

(7) Local minimum value is 0 at $x = 1$.   (8) Local minimum value is 0 at $x = 0$.

(9) Local maximum value is $\frac{1}{3}$ at $x = 3$ and local minimum value is $-\frac{1}{2}$ at $x = -2$.

(10) Local minimum value is $\sqrt{3}$ at $x = \frac{\pi}{3}$.   (11) Local maximum value is $\frac{1}{4e}$ at $x = e^{\frac{1}{4}}$.

#15 (1) 1) False 2) True 3) False   (2) 1) True   2) True 3) False 4) False

(3) 1) True 2) False 3) False 4) False   (4) 1) False 2) True 3) False

(5) 1) False 2) True 3) False   (6) 1) False 2) True 3) True

#16 (1) Local minimum value is 0 at $x = 0$ and local maximum value is $\frac{4}{e^2}$ at $x = 2$.

(2) Local minimum value is -2 at $x = \frac{4\pi}{3}$ and local maximum value is 2 at $x = \frac{\pi}{3}$.

(3) Local minimum value is $-\frac{\pi}{2} + 1$ at $x = -\frac{\pi}{4}$ and local maximum value is $\frac{\pi}{2} - 1$ at $x = \frac{\pi}{4}$.

#17 (1) 1) $a > 3$ or $a < 0$ 2) $0 \leq a \leq 3$ (2) $a = 10$, Local minimum value is 8.

(3) $a = -1, \ b = -1$

(4) 1) $a = -2, \ b = 9$

2) Local maximum value at $x = 1$ is $4 + c$ and local minimum value at $x = 3$ is c.

3) $c = 6$, Local minimum value is 6.

(5) $a = 0$ or $a \geq \frac{3}{4}$ (6) $a < -\frac{5}{2}$ (7) $-2 < a < -1$ (8) $= \sqrt{10}, b = -\sqrt{10}$

(9) 0   (10) $0 < a < \frac{1}{4}$ (11) 1) $\frac{\pi}{3} < \theta \leq \frac{\pi}{2}$ 2) $0 \leq \theta \leq \frac{\pi}{3}$ (12) $|a| < 2$

(13) $= -1$, Local maximum value is $\frac{\sqrt{2}}{2} e^{-\frac{\pi}{4}}$

#18 $f(x) = \frac{2}{3}x^3 - 2x + 2$

#19 (1) $(0, 4)$ and $(2, -12)$ (2) $\left(\dfrac{e}{a}, 1\right)$

#20 (1) $y = x - \dfrac{\pi}{4} + \dfrac{1}{2}$ (2) $y = -\dfrac{1}{e^2}x + \dfrac{4}{e^2}$

#21 $a = b = 2$

#22 2

#23 (1) Maximum value: 4 at $x = 3$,   Minimum value: $-\dfrac{1}{2}$ at $x = 0$

   (2) Maximum value: 50 at $x = 5$, Minimum value: 0 at $x = 0$

   (3) Maximum value: 1 at $x = 0$, Minimum value: $-1 - \sqrt{2}$ at $x = -1 - \sqrt{2}$

   (4) Maximum value: $3\sqrt{2} - 3$ at $x = \dfrac{3}{\sqrt{2}}$, Minimum value: -6 at $x = -3$

   (5) Maximum value: 1 at $x = 1$, Minimum value: $-\dfrac{1}{3}$ at $x = -1$

   (6) Maximum value: 3 at $t = 0$ $(x = 0, \pi, 2\pi)$, Minimum value: -2 at $t = -1$ $\left(x = \dfrac{3\pi}{2}\right)$

   (7) Maximum value: $\dfrac{3\sqrt{3}}{4}$ at $x = \dfrac{\pi}{3}$, Minimum value: 0 at $x = \pi,\ x = 0$

   (8) Maximum value: $\dfrac{13}{8}$ at $t = \dfrac{1}{2}$ $\left(x = \dfrac{\pi}{6},\ x = \dfrac{5\pi}{6}\right)$, Minimum value: $-\dfrac{7}{4}$ at $t = -1\left(x = \dfrac{3\pi}{2}\right)$

   (9) Maximum value: $\sqrt{3} + \dfrac{\pi}{3}$ at $x = \dfrac{\pi}{3}$, Minimum value: $-\sqrt{3} + \pi$ at $x = \pi$

   (10) Maximum value: $\sqrt{2}$ at $t = \dfrac{1}{\sqrt{2}}$ $\left(x = \dfrac{\pi}{4},\ x = \dfrac{3\pi}{4}\right)$, Minimum value: -1 at $t = -1\left(x = \dfrac{3\pi}{2}\right)$

   (11) Maximum value: 6 at $t = 0$, Minimum value: $-2\sqrt{2} - 6$ at $t = -\sqrt{2}$

   (12) Maximum value: $-1 + \sqrt{2}$, Minimum value: $-1 - \sqrt{2}$

   (13) Maximum value: e^3 at $x = 3$, Minimum value: $-4e^2$ at $x = 2$

   (14) Maximum value: $\sqrt{2}e^2$ at $x = 2$,   Minimum value: 0 at $x = -\sqrt{6},\ x = \sqrt{6}$

#24 (1) $a = \dfrac{2}{3}$, $b = 3$ (2) 8   (3) $a = \dfrac{\pi}{4}$, $b = \dfrac{\pi}{2}$ (4) $a = 0$, $b = 2$ (5) $a = 1 + e^{-4}$

#25 (1) 1) $-4 < a < 4$ 2) $a = 4$ or $a = -4$ 3) $a > 4$ or $a < -4$

   (2) $-80 < a < 0$ (3) $a < -50$ (4) $-32 < a < -5$ (5) $-\sqrt{3} < a < \sqrt{3}$ (6) $0 < a < 1$

   (7) $-2 < a < 2$ (8) $0 < a < \dfrac{1}{e}$ (9) $-\pi \le a \le \sqrt{3} - \dfrac{\pi}{3}$ (10) $a > 3$ or $a < 2$

   (11) 1) $-15 < a < 17$ 2) $a = 17$ or $x = -15$ 3) $a > 17$ or $a < -15$

(12) $1 < a < 3$ (13) $a < -\frac{2}{3}$ (14) $0 < a < \frac{e}{2}$ (15) $0 \le a \le 1$

#26 (1) $a < -1$ (2) $a = -e^2$ (3) $\frac{1}{e}$ (4) -34

#27 15 (seconds)

#28 (1) At $t = 6$, $V(t) = 4$ and $A = -16$; At $t = 7$, $V(t) = -12$ and $A = -16$

(2) 312.5 feet

(3) At the end of 2.5 sec of motion, the object is at a height of 200 feet is rising since $V > 0$.
At the end of 10 sec of motion, it is at the same height but is falling since $V < 0$.

#29 At $t = 3$

#30 $8 < m < 10$

#31 $8\sqrt{10}$ (miles/hour)

#32 (1) At 2.8 miles/hour   (2) At 6 miles/hour   (3) When $t = 1.28$ hour, 9.6 miles apart.

#33 5

#34 $t = 2$

#35 (1) 2   (2) $-\frac{\pi}{3}$

#36 $\theta = 6$, $V = \frac{8}{3}$ rad/sec, $A = \frac{2}{3}$ rad/sec^2

#37 (1) $x = r\cos\omega t$, $y = r\sin\omega t$

(2) $V(t) = \langle -r\omega\sin\omega t, r\omega\cos\omega t \rangle$, $A(t) = \langle -r\omega^2\cos\omega t, -r\omega^2\sin\omega t \rangle$

(3) $r\omega$

(4) See Solutions Manual.

#38 (1) $\frac{1}{\pi}$ feet/min   (2) $\frac{2}{5}$ feet^2/min   (3) $\frac{\sqrt{17}}{4}$ feet^2/min

Chapter 5. The Indefinite Integral

#1 (1) $a = 4$, $b = 1$, $c = -4$ (2) $a = 8$, $b = 3$, $c = 2$

#2 (1) $\frac{1}{6}x^6 + C$ (2) $x^3 + 3x^2 + C$ (3) $\frac{2}{3}x^3 + \frac{3}{2}x^2 - 4x + C$ (4) $2\log|x| + \frac{3}{x} + C$

(5) $2\log|x| - x + C$ (6) $x^3 - \frac{2}{7}x^{\frac{7}{2}} + C$ (7) $\frac{1}{4}x^4 + \frac{1}{3}x^3 - x^2 + C$ (8) $2\theta + C$

(9) $-x + C$ (10) $\frac{1}{3}x^3 - \frac{1}{2}x^2 + x + C$ (11) $\frac{2}{5}x^2\sqrt{x} - \frac{4}{3}x\sqrt{x} - 6\sqrt{x} + C$

(12) $\frac{1}{4}x^4 - \frac{3}{2}x^2 + 3\log|x| + \frac{1}{2x^2} + C$ (13) $x - 8\sqrt{x} + 4\log|x| + C$ (14) $\frac{1}{3}x^3 + \frac{1}{2}x^2 + x + C$

#3 (1) $4x$ (2) $4x^2 - 1$ (3) $f(x) = 2x - 1$ and $g(x) = 2x + 1$

#4 (1) $f(x) = \frac{2}{3}x^3 - \frac{3}{2}x^2 + x + \frac{11}{6}$

(2) $\int f(x)dx = \frac{1}{2}x^3 + 2x^2 + C_1 x + C_2$, The coefficient of x^2 is 2.

(3) $f(x) = \frac{2}{7}x^{\frac{7}{2}} - \frac{1}{2}x^2 + 3x + 4$ (4) $x = \frac{1}{2}$ (5) 6   (6) $f(x) = \log x + 3$ (7) 1

(8) 112   (9) The local maximum value is $\frac{8\sqrt{6}}{9} + 1$ and local minimum value is $-\frac{8\sqrt{6}}{9} + 1$

(10) $a = 5$, $f\left(\frac{5}{3}\right) = \frac{104}{27}$

(11) $f(x) = \frac{1}{3}x^3 - 2x^2 + 3x + \frac{11}{3}$ The local minimum value is $f(3) = \frac{11}{3}$

(12) $\frac{125}{6}$ (13) $-\frac{4}{9}$ (14) $k < 4$ (15) $-2e - 2$ (16) $\frac{16}{3}$ (17) $e^x + 2e$

(18) 6   (19) 9

#5 (1) $3x + \cos x + C$ (2) $-\cot x - x + C$ (3) $\tan x + C$ (4) $x + \sin x + C$ (5) $\sec x + C$

(6) $-\csc x + C$ (7) $\log\left|\frac{\sec x + \tan x}{\csc x + \cot x}\right| + C$ (8) $\log\frac{(\sec x + \tan x)^4}{|\cos x|} + C$ (9) $\frac{1}{2}x - \frac{1}{2}\sin x + C$

(10) $\tan x - \cot x + C$ (11) $2e^x + 3\cot x + C$ (12) $\frac{4^x}{\log 4} - \frac{2^x}{\log 2} + x + C$

(13) $\tan x - \log|x| + C$

#6 (1) $f(x) = -\frac{1}{2}\cos x - 1$ (2) $2e$ (3) $y = \frac{1}{12}x^4 - \frac{1}{2}x^2 + \frac{1}{6}x - \frac{3}{4}$ (4) $y = 2e^{x^3}$

#7 (1) $\sin x$ (2) $e^{-x}(-\cos x + 2)$

#8 (1) $\frac{1}{3}(x^3+2)^3+C$ (2) $\frac{2}{9}(x^3+2)^{\frac{3}{2}}+C$ (3) $\frac{1}{8}(x^2+2x+1)^4+C$ (4) $-\frac{4}{3(x^3+4)^2}+C$

(5) $\frac{4}{9}(x^3+1)^{\frac{3}{4}}+C$ (6) $-\frac{2}{9}(1-3x^2)^{\frac{3}{2}}+C$ (7) $\frac{1}{3}(x^2+1)\sqrt{x^2+1}+C$

(8) $\frac{2}{3}(x-5)\sqrt{x+1}+C$ (9) $\frac{3}{4}(x^2+8x)^{\frac{2}{3}}+C$ (10) $-\frac{180}{\pi}\cos x^\circ+C$

(12) $\frac{1}{2}\sin^2 x+C$ or $-\frac{1}{2}\cos^2 x+C$ (12) $-\frac{2}{3}\cos^3 x+\cos x+C$ (13) $\frac{1}{3}\sin^3 x+C$

(14) $x-\frac{1}{2}\cos 2x+C$ (15) $\frac{1}{4}\sin 2x+\frac{1}{2}x+C$ (16) $\frac{1}{2}x-\frac{1}{4}\sin 2x+C$

(17) $-\frac{1}{10}\cos 5x+\frac{1}{2}\cos x+C$ (18) $\frac{1}{4}\sin 2x+\frac{1}{12}\sin 6x+C$ (19) $-\frac{1}{5}(1+\cos x)^5+C$

(20) $-\log|\cos x|+C$ or $\log|\sec x|+C$ (21) $\log|\tan x|+C$ (22) $-\cot x+\csc x+C$

(23) $\log\left|\tan\frac{x}{2}\right|+C$ (24) $-2\cos\sqrt{x}+C$ (25) $\frac{1}{5}(e^x+1)^5+C$ (26) $-\frac{1}{4}e^{-4x}+C$

(27) $\frac{1}{2}e^{x^2}+C$ (28) $2\log(e^x+1)-x+C$ (29) $\sin e^x+C$ (30) $-\frac{1}{6}e^{3\cos 2x}+C$

(31) $\frac{a^{2x}}{2\log a}+C$ (32) $\frac{5^{4x+3}}{4\log 5}+C$

#9 (1) $\log|x+2|+C$ (2) $x+\log|x+1|+C$ (3) $\frac{1}{2}x^2+x+2\log|x-1|+C$

(4) $\frac{1}{2}\log|x^2-1|+C$ (5) $-\frac{1}{6}\log|1-2x^3|+C$ (6) $\log|x^2+x+1|+C$

(7) $\log|2x-1|-\log|x+1|+C$ (8) $\frac{1}{40}\log\left|\frac{5+4y}{5-4y}\right|+C$ (9) $\frac{1}{24}\log\left|\frac{3x-4}{3x+4}\right|+C$

(10) $-2\log|x-1|+3\log|x-2|+C$ (11) $\frac{2}{3}(x+1)\sqrt{x+1}-\frac{2}{3}x\sqrt{x}+C$

(12) $\frac{3}{4}\sqrt[3]{(2x+1)^2}+C$ (13) $-\log|\cos x|+C$ (14) $\log|x+\sin x|+C$ (15) $\log|\log x|+C$

(16) $x-\log(e^x+1)+C$

#10 (1) $\frac{1}{3}xe^{3x}-\frac{1}{9}e^{3x}+C$ (2) $\frac{1}{4}(2x^2-2x+1)e^{2x}+C$ (3) $\frac{1}{2}x\sin 2x+\frac{1}{4}\cos 2x+C$

(4) $=(2-x^2)\cos x+2x\sin x+C$ (5) $-\frac{1}{4}\sin 2x+\frac{1}{2}x+C$ (6) $\frac{1}{3}x^3\log x-\frac{1}{9}x^3+C$

(7) $(x+2)\log(x+2)-x+C$ (8) $\frac{2}{3}x(1+x)^{\frac{3}{2}}-\frac{4}{15}(1+x)^{\frac{5}{2}}+C$

(9) $2\sqrt{x-1}\log(x-1)-4\sqrt{x-1}+C$

#11 (1) 1) $\frac{1}{2}e^x(\cos x + \sin x)$ 2) $\frac{\pi}{2}$ (2) $\frac{1}{2}x^2 \log x - \frac{1}{4}x^2 + \frac{13}{4}$ (3) $f(x) = e^{x+2}$

(4) $\frac{5}{2}$ (5) e^2 (6) $\frac{1}{4}$ (7) $\frac{3}{2}$ (8) 12   (9) 128

(10) $\frac{225}{2}$ feet.   (11) 1) 675 feet   2) 9 seconds   3) 180 feet/sec

Chapter 6. The Definite Integral

#1 (1) 26 (2) −15 (3) $\frac{2}{3}$ (4) 0 (5) 1 (6) $-\frac{1}{e^2}+1$ (7) −2 (8) $\frac{8}{3}$ (9) $\frac{3}{2}+2\log 2$

(10) $2\log 2 - \log 3$ (11) $\frac{1}{2}\log\frac{5}{2}$ (12) $\sqrt{3}-\frac{\pi}{3}$ (13) $\frac{1}{4}(e^{2\pi}-\pi-1)$ (14) $\frac{1}{2}\left(1-\frac{1}{e}\right)$

(15) $4\sqrt{6}$ (16) −5 (17) $-\frac{29}{3}$ (18) $\frac{4}{3}\sqrt{2}$ (19) $-1+\log 2$ (20) $\frac{3}{2}+3\log 2$

(21) $\log 2 - \log 3$ (22) 4 (23) 2π (24) $\frac{1}{8}$ (25) 0 (26) 1 (27) $-\frac{1}{2}$ (28) $\log 2 + \frac{1}{2}$

#2 (1) $\frac{5}{3}$ (2) $\frac{3}{2}$ (3) $\frac{3\sqrt{3}}{4\pi}$ (4) 1 (5) $\log 2$ (6) $\frac{1}{k+1}$ (7) $2\log 2 - 1$ (8) 1

(9) $2\log 2 - 1$ (10) 1 (11) $\frac{1}{3}$ (12) $\frac{\sqrt{3}\pi}{18}$ (13) $\frac{15}{4}$

#3 (1) 0 (2) $-A$ (3) $A+B$ (4) $A+B+2C$ (5) $\sqrt{2}C-(A+B)$

#4 (1) $\frac{3}{2}$ (2) 22 (3) 8 (4) 1 (5) 14 (6) 2 (7) $\log\sqrt[3]{\frac{9}{2}}$ (8) 2

#5 See Solutions Manual.

#6 (1) $\frac{1}{2}$ (2) $\frac{5}{2}$ (3) 1   **#7** (1) $\frac{14}{3}$ (2) $4\frac{5}{6}$ (3) $6\frac{1}{3}$

#8 (1) 1 (2) $27e^3$ (3) $\sqrt{5}$ (4) 7 (5) 4 (6) 0 (7) $3e$ (8) $\frac{6}{5}$ (9) $k!$

#9 (1) 2 (2) $6\log 3 + e - 8$ (3) 42 (4) 13

(5) i) If $0 \le a \le 2$, then $a^2 - 2a + 2$ ii) If $a \ge 2$, then $2a - 2$

(6) $\frac{1}{3}(7-4\sqrt{2})$ (7) $4\log\frac{8}{3}-1$ (8) $2\sqrt{2}$ (9) 1 (10) $\frac{4}{3\log 3}-\frac{1}{2\log 2}$

#10 (1) $x^4 + 3x^2 - 5x + 2$ (2) $-\cot^2 x \sin x$ (3) $4x^3 - 2x$ (4) $-3\sqrt{9x^2+3}$

#11 (1) $f(x) = 3x - \frac{27}{4}$ (2) $f(x) = \sin x - 1$ (3) $f(x) = e^x + 4x$ (4) $f(x) = e^x + 2$

(5) $f(x) = x - 4$ and $g(x) = 3x - \frac{11}{2}$ (6) $f(x) = 6x^2 + 1$ (7) $f(x) = e^x$

(8) $f(x) = \frac{1}{2}\left(x - \frac{1}{2}\sin 2x\right) - \frac{3}{2}\pi$

(9) $f(x) = 3x^2 + 2x, \quad g(x) = 3x^2 - 4x + 2, \quad a = -1, \ b = -3$

(10) $f(x) = 2e^{2x} - 2e^x, \quad a = \log 2$ (11) $f(x) = e^x, \quad a = b = -1$

(12) $a = \log 1 = 0$ or $a = \log 3, \quad f(x) = 2e^{2x} - 4e^x$

(13) $a = -2, \quad f(x) = \pi\cos\pi x - 4\pi\sin 2\pi x$

#12 (1) $x = 3$ (2) $x = 3$

#13 (1) $3e - 1$

    (2) 1) $f(x)$ has a local maximum value at $x = \pi$ and local minimum value at $x = 2\pi$.

      2) 0

   (3) $a = -3$

#14 (1) $e^e(e^2 - 2e + 2) - e$ (2) π (3) $\frac{52}{9}$ (4) $\log 5$ (5) 1 (6) $\frac{1}{4}$ (7) $\frac{1}{2}$ (8) -1

    (9) π (10) $1 - \log 2$ (11) $\frac{1}{4}$ (12) $2(e^2 - e)$ (13) $\log 2 - \frac{1}{2}$ (14) $\frac{1}{4}(\log 2)^2$

    (15) $\log\frac{2e}{e+1}$ (16) $\frac{1}{2}\log\frac{3(e-1)}{e+1}$ (17) $\frac{a^2\pi}{4}$ (18) $\frac{\pi}{4a}$

#15 (1) 2 (2) $-\frac{2}{3}$ (3) $16\log 4 - 12$ (4) $\frac{1}{6}(e^4 - e)$ (5) 2 (6) $4 \le \int_1^3 f(x)dx \le 10$

    (7) 8 (8) $-A + 2B$

#16 (1) -16 (2) 0 (3) 0 (4) 4 (5) 0 (6) $-\frac{16}{3}$ **#17** 8

#18 (1) $\frac{1}{4}$ (2) $\frac{1}{4}(e^2 - 1)^2$ (3) $\frac{5}{2}$ (4) 1 (5) 7 (6) $\frac{2}{3\pi}$ (7) 2 (8) $\frac{100}{101}$

    (9) 12 (10) $e - 2$ (11) $e - 2$ (12) $a = -1$ (13) $a = \pm\frac{2\sqrt{3}}{3}$ (14) $-\frac{2}{e}$ (15) $-e$

    (16) $\frac{\pi}{4}$ (17) $g\left(\frac{3}{2}\right)$ (18) $-\frac{1}{4}$ (19) -1 (20) -18 (21) $a^2 - 4b < \frac{1}{3}$ (22) 18

    (23) 30 (24) -10 (25) 2 (26) -2 (27) 2

Chapter 7. Applications of the Integral

#1 (1) $\frac{8}{3}$ (2) $\frac{23}{4}$ (3) $\frac{4}{3}$ (4) $\frac{1}{2}$ (5) 1   (6) 2   (7) $\frac{32}{3}$ (8) 8   (9) $\frac{512}{25}$ (10) 1   (11) $\frac{4}{3}$

(12) 4π (13) $\log 2$ (14) 2   (15) $\frac{1}{6}$ (16) $e + \frac{1}{e} - 2$

#2 (1) 2   (2) $\frac{\pi}{2}$ (3) 32   (4) $-2\sqrt{3}$ (5) $\frac{1}{4}$ (6) $\sqrt[3]{\left(\frac{15}{4}\right)^2}$ (7) $a(e^a - 1)$ (8) $\frac{8}{3}$ (9) $\frac{9}{2}$

(10) $\sin a + a \cos a$ (11) $\frac{1}{6}$ (12) $\frac{4}{3}$

#3 (1) $\frac{9}{2}$ (2) $\log\sqrt{2}$ (3) $\frac{9}{2}$ (4) 8   (5) $\frac{37}{12}$ (6) $\frac{7}{15}$ (7) $\frac{8}{3}\sqrt{3}$ (8) $\frac{11}{4}$ (9) $\frac{13}{3}$ (10) $\frac{1}{6}$

(11) $-\frac{5}{e^2} + 1$ (12) $\frac{9\pi}{2} - 1$ (13) $\frac{5}{2}$ (14) 9   (15) $\frac{9}{2}$ (16) $\frac{1}{3}$ (17) $\frac{32}{3}$ (18) $\frac{2}{3}$

(19) $6 - 3\log 3$ (20) $\frac{1}{2}e - 1$ (21) $\frac{e}{2} - 1$ (22) $\frac{\pi}{3}$ (23) $\frac{1}{n(n+1)}$ (24) $1 - \frac{2}{n+1}$

#4 (1) $\frac{35}{6}$ (2) $\frac{27}{4}$ (3) 2   (4) 108   (5) $\frac{4}{3}$ (6) $\frac{1}{4} + \frac{1}{e^2} - \frac{1}{4e^4}$ (7) $\frac{2}{3}\sqrt{e} - 1$ (8) $\frac{1}{2}e - \frac{1}{e^2}$

(9) $ab\pi$ (10) $\frac{8}{3}$ (11) $\frac{14}{3}$ (12) 18   (13) $A = \frac{5}{9}$, $B = \frac{1}{9}$

#5 (1) $\frac{128}{3}\pi$ (2) $\frac{128}{3}\pi$

#6 (1) 8   (2) $\frac{8}{15}\sqrt{2}$ (3) 20π (4) $\frac{16}{15}\sqrt{3}$

#7 (1) $\frac{\pi^2}{8}$ (2) $\frac{\sqrt{3}}{8}\pi$

#8 (1) 8   (2) $\frac{256}{15}$ (3) $\frac{32}{3}$

#9 18

#10 (1) $\frac{16}{15}\pi$ (2) $\frac{56}{15}\pi$ (3) $\frac{1}{2}\pi$ (4) $\frac{2}{3}\pi$ (5) $\frac{8\pi}{3}$ (6) $\frac{32}{15}\pi$ (7) $\pi(3\pi + 8)$

(8) $\frac{\pi}{2}(e^2 - 4e + 5)$ (9) $\pi(e - 2)$ (10) $\frac{108}{5}\pi$ (11) $\frac{16}{3}\pi$ (12) $\frac{\pi}{2}\left(1 - \frac{1}{e^2}\right)$

(13) $\frac{\pi}{2}\left(e^2 + 4 - \frac{1}{e^2}\right)$ (14) $\pi\left(1 - \frac{\pi}{4}\right)$ (15) $\frac{3\sqrt{3}}{16}\pi$ (16) $3\pi^2$ (17) π^2 (18) $\frac{20}{3}\pi$

(19) $\frac{16}{3}\pi$ (20) $32\pi^2$ (21) $\frac{64}{3}\pi$

#11 (1) $\frac{1}{2}\pi$　(2) $\frac{8}{15}\pi$　(3) $\frac{1}{5}\pi$　(4) $\frac{1}{4}\pi$　(5) $\frac{1}{3}\pi$　(6) $\frac{16}{3}\pi$　(7) $\frac{2}{15}\pi$　(8) $8\pi^2$

(9) $\frac{\pi}{2}(e^2 - 4e + 5)$　(10) $\frac{3}{2}\pi$　(11) $\pi\left(4\log 2 - \frac{5}{2}\right)$　(12) $\pi(e-2)$　(13) $\pi\left(e - \frac{7}{3}\right)$

(14) $\pi\left(1 - \frac{2}{e}\right)$

#12 (1) $a = \sqrt{6}$　(2) $a = 4$　(3) $\frac{256}{315}\pi$　(4) $a = 2$　(5) $a = 1$

#13 (1) (2) See Solutions Manual.

#14 (1) $\frac{48\sqrt{2}}{5}\pi$　(2) $\frac{32\sqrt{2}}{5}\pi$

#15 (1) $\frac{11}{2}$　(2) $e - 1$　(3) $2\left(\frac{1}{e} - \frac{1}{e^2}\right)$

#16 (1) 25 feet　(2) 30 feet　(3) 25 feet

#17 $\langle e, -\frac{1}{e} \rangle$

#18 900 feet

#19 972π

#20 300 feet^3

#21 The velocity more than 20 feet/sec

#22 52 feet

#23 (1) $\sqrt{5}$　(2) 13　(3) 8　(4) $\frac{2}{27}\left(11\sqrt{22} - 4\right)$　(5) $e - \frac{1}{e}$　(6) $e - \frac{1}{e}$　(7) $\sqrt{2}\left(1 - \frac{1}{e^{\pi}}\right)$

(8) $\frac{1}{2}\log 3$

#24 $\frac{1}{4e^4} - e^4$

Index

A

B

C

T

U

V

W

Printed in Great Britain
by Amazon